# CIVIL NUCLEAR ENERGY
## Fuel of the Future or
## Relic of the Past?

Malcolm C. Grimston
and Peter Beck

## THE ROYAL INSTITUTE OF
## INTERNATIONAL AFFAIRS
**Energy and Environment Programme**

Published in Great Britain in 2000
by the Royal Institute of International Affairs,
Chatham House, 10 St James's Square, London SW1Y 4LE
(Charity Registration No. 208 223)

Distributed worldwide by the Brookings Institution,
1775 Massachusetts Avenue, NW, Washington, DC 20036-2188

ISBN 1 86203 128 2

Cover design by Matthew Link
Printed and bound in Great Britain by the Chameleon Press Ltd

*The Royal Institute of International Affairs is an independent body which
promotes the rigorous study of international questions and does not express
opinions of its own. The opinions expressed in this publication are the
responsibility of the authors.*

# CONTENTS

Foreword    v
About the authors    viii

1   Introduction    1
   1.1   The problem    1
   1.2   The project    4
   1.3   Starting points    4
   1.4   The position paper    15
   1.5   Alternative approaches    16

2   Issues over which the nuclear industry has
relatively little influence    18
   2.1   Introduction    18
   2.2   Energy choices and the 'need' for nuclear power    19
   2.3   Global climate change    37
   2.4   Low-level radiation    45
   2.5   Regulatory issues    49

3   Issues over which the nuclear industry has
major influence    53
   3.1   Introduction    53
   3.2   Keeping the nuclear option open    55
   3.3   The relative economics of nuclear power    58
   3.4   Nuclear research and development    68
   3.5   Skills required if nuclear power is phased out    81

| 4 | Issues over which the nuclear industry has partial influence | 86 |
|---|---|---|
| | 4.1 Introduction | 86 |
| | 4.2 Nuclear proliferation | 86 |
| | 4.3 Public perception, politics and decision-making | 90 |
| | 4.4 Waste management | 98 |
| | 4.5 Reprocessing | 106 |
| | 4.6 Safety of nuclear installations | 109 |
| 5 | Conclusions | 116 |
| Appendix 1: The WEC/IIASA scenarios | | 119 |
| Appendix 2: Global status of the nuclear industry | | 122 |
| References | | 124 |

# FOREWORD

Recent events – high oil prices, extreme weather conditions, and the Conference of Parties in The Hague to take forward the implementation of the Kyoto Protocol – should perhaps have thrown the spotlight on civil nuclear power. It is, after all, the most significant source of energy currently available which does not rely on fossil fuels and which emits no carbon dioxide. Yet, at a time when these benefits should be of ever growing importance, there seems to be a real possibility that, far from growing, nuclear power may in a number of countries be heading for a possibly terminal decline.

There are many reasons – economic, environmental, concerns about safety and proliferation – for doubting whether nuclear power has a long-term future. But there are also those who argue that it could have a significant role in mitigating climate change and providing secure and reliable energy for the twenty-first century. Positions on both sides of the argument tend to be polarized and deeply entrenched, making it difficult to find common ground for debate. Many, including some in government, have apparently preferred to avoid controversy by not raising the question at all. Yet, whether or not it is right to seek to retain the nuclear option, it is surely wrong to let the option expire by default simply because the issues are too difficult to address; to avoid debate because debate might be contentious; and to ignore the issues because objective information is difficult to obtain.

It is with the aim of promoting a better-informed debate that the Royal Institute of International Affairs has set up this enquiry into the future of civil nuclear energy, following a more general study of the subject in 1994. The aim of the enquiry is not to make a case for or against nuclear power, but to clarify the issues; to set out clearly and objectively the main arguments for and against nuclear power; and to highlight the questions that arise from this analysis. Phase 1 of the study, with which the present report is concerned, surveys the main arguments. Phase 2, which is now under way, will investigate key issues in more detail. A key output of Phase 2 will be a programme of disseminating the results to decision-makers.

The study was overseen by Peter Beck, a long-standing adviser to the Energy and Environment Programme, and undertaken by Malcolm Grimston. Peter is used to taking the long view, having a background in scenario planning at Shell and as a past President of the European Strategic Planning Federation. Malcolm has worked in and observed the energy field for thirteen years, eight with the UK Atomic Energy Authority and five with Imperial College, London, where he is still a Visiting Fellow.

The study has been subject to extensive peer review and scrutiny, including a special workshop on the issue. It is a tribute to the authors that they have produced a report that is not only readable and comprehensible, even for non-experts, but that is also seen by the participants in this process, from both sides, as having maintained a scrupulous neutrality and objectivity.

I very much welcome this important study and hope that it will succeed in its aim of helping stimulate a lively, well-informed debate on nuclear power and its future. We cannot form a clear picture of future global energy supplies, or of the world's response to climate change, without understanding what role nuclear power might play.

Malcolm Keay
*Deputy Head*
*Energy and Environment Programme*

## ABOUT THE AUTHORS

**Malcolm Grimston** is a Senior Research Fellow in the Energy and Environment Programme at the Royal Institute of International Affairs. After reading Natural Sciences at Magdalene College, Cambridge, he taught chemistry for seven years. In 1987 he joined the UK Atomic Energy Authority as an information officer. In 1995 he was appointed as a Senior Research Fellow at the Centre for Environmental Technology, Imperial College, London, working in the Energy and Environment Policy Group, and he retains Visiting status at the college.

Malcolm Grimston's publications include *Coal as an Energy Source*, for the International Energy Agency, *Chernobyl and Bhopal Ten Years On: Comparisons and Contrasts* and *Leukaemia and Nuclear Establishments – Fifteen Years of Research*, as well as numerous journal articles. He is a regular media contributor on energy and nuclear matters.

**Peter W. Beck** is an Associate Fellow in the Energy and Environment Programme at the RIIA where he is currently involved in the research project on 'The Worldwide Future of Civil Nuclear Energy'. He is an adviser to the Oxford Institute for Energy Studies, and has been a member of an international working group on setting up an internationally monitored storage and management regime for spent reactor fuel and plutonium from civil nuclear power plants.

Peter Beck has been active in matters relating to business planning and strategy, having been Planning Director of Shell UK Ltd during a career of nearly forty years with the Shell Group, Chairman of the Strategic Planning Society in UK and subsequently President of the European Strategic Planning Federation. He has published and lectured widely on subjects related to the future of nuclear power, the practice of strategic planning, energy policy and the futility of much of energy forecasting.

# 1 INTRODUCTION

## 1.1 The problem

World energy use has grown consistently for many decades, and is expected to continue to grow as population increases and poorer countries embrace industrialization. Much of the projected growth will therefore take place in currently less developed countries. Plentiful energy supplies, especially electricity, lie at the basis of modern lifestyles in the developed world, lifestyles to which many developing countries aspire. Global emissions of greenhouse gases, especially carbon dioxide, have also grown significantly, and, even assuming success of the Kyoto Protocol, are on an upward trend.

Although there is some confidence that demand for energy will increase – bearing in mind that demand is often difficult to forecast – projections of which sources will be used to supply energy in the various regions of the world are much more problematic.

Matters are further complicated by the enormous timescales involved in energy. From first examination of a new technique to widespread commercial application can take a number of years. Expanding the world's gas pipelines, installing significant solar capacity, constructing a large programme of nuclear power stations or wind generators, all are likely to take decades. History suggests that commercial exploitation of new technologies is generally rather slower than is regarded as feasible or desirable by their promoters. In certain circumstances, e.g. the 'dash for gas' in the UK in the 1990s, it can proceed rapidly if there is clear market demand, but only if an appropriate infrastructure is in place.

It is essential, then, that cool objective consideration is given to all possible strategies to provide for future energy requirements and reduce emissions of greenhouse gases.

Nuclear power represents one such possibility. However, at present its future is a subject of acrimonious dispute in a number of countries. Several possible reasons have been adduced for the particularly vitriolic nature of the civil nuclear debate in these countries. They include:

- the perceived link between nuclear power and nuclear weapons;

- concerns over long-term risks and management of nuclear wastes;

- perception of secrecy in decision-making;

- perception of a culture of cover-up and dishonesty;

- the apparently 'imposed' nature of nuclear risks upon many people;

- the fact that nuclear radiation is unfamiliar to many people, and undetectable by human sensory organs;

- the apparent potential for nuclear power to cause damage to vast numbers of people and to future generations, especially through large accidents;

- the perceived irreversibility of many nuclear decisions;

- the politically centralizing nature of nuclear technology, in contrast, say, to renewables;
- the absence of a strongly perceived link between the activity and the benefit.

It is noticeable that in countries such as France where nuclear power is well established, and regions of other countries which act as host to nuclear stations and hence where the 'fear of the unknown' is lowest, public confidence in nuclear power tends to be highest.

The poor state of public confidence in nuclear power in some countries, coupled with questions over its economics in a free market, have led to some speculation that it may not have a future at all in the medium or longer term. As the dispute rages, especially over issues such as the management of radioactive waste, safety of installations and proliferation of nuclear weapons, so reaching decisions becomes ever more difficult. As in many areas of dispute, the political attractions of doing nothing are considerable. Even if it is assumed that decisions will increasingly be taken by organizations in the private sector, the stance taken by governments will remain crucial.

Without a better understanding of the issues, and in particular the reasons for the disputes, important public policy decisions may not be taken, simply by default. However, as the world moves towards the tenth anniversary of the Rio Convention in 2002, it is clear that major policy decisions may well be needed very shortly if climate change, for example, is to be restricted to manageable proportions. Whether or not nuclear power could or should have a role in mitigating climate change, in improving the security of world energy supplies or in reducing energy costs is an open question, but an important one. It is unlikely that abandoning this or any other technological option, simply because it is too difficult to cut through levels of emotion and vitriol on both sides, would result in the best use of potential resources.

## 1.2 The project

Following on a more general study of the subject (Beck, 1994), the Royal Institute of International Affairs has established an enquiry into the future of civil nuclear energy. This paper is one of the main outputs of Phase 1, which ran from October 1999 to September 2000. The aims of this phase were to clarify the main areas of agreement and dispute within the nuclear field; to identify which of these have been relatively well researched and those which have been relatively neglected; and to propose a programme of research for Phase 2. Phase 2 will involve the production of a series of research or review reports considering the major issues in greater depth.

## 1.3 Starting points

In the field of energy there are very few global generalizations to be made. In some countries in the developed world energy use is practically static, while it is increasing at high rates in some developing and newly industrialized countries – it more than doubled in South Korea and Thailand during the 1990s (BP Amoco,

2000). In some developed countries, such as Japan, security of supply is a major driving factor behind energy strategy, while in others, such as the UK, security of supply is not considered problematic for the foreseeable future. Some countries, such as Sweden, Norway and the Netherlands, have traditionally taken global environmental issues very seriously, while for other countries factors such as providing energy for large populations living in poverty have understandably taken higher priority. In developing countries such as China and India, nuclear power, including reprocessing, is seen as a central component of strategies to provide secure and environmentally acceptable electricity production, while in Germany and Sweden it is officially planned to phase out nuclear power before the end of the economic life of the stations, and indeed Italy has already followed this course. In some countries nuclear power appears popular, in others it is not a major public issue, in others still it is a focus of discontent. In some countries the economics of large plants appears favourable, in others no nuclear construction is at present contemplated by private-sector funding institutions. Of course the current situation in any country could change in the medium or long term, but wide global diversity may well persist.

### 1.3.1 The nuclear debate

As noted above, the debate over nuclear power in some developed countries at least has been characterized by unusually entrenched positions on both sides. The issue has a tone not usually found in discussions over the future use of coal or natural gas, for example, or over wind power or solar power, despite the considerable advantages and disadvantages associated with these fuels. Discussions in these areas tend to be rather more matter of fact. Even disputes over large-scale hydro-power do not seem to provoke quite the same level of denigration of opponents. However, a similar flavour can be found in debates over genetically modified organisms in food, animal experimentation and, recently, global capitalism in general.

To embark on a project designed to increase understanding about the areas of dispute between the 'sides', insofar as one can characterize such a complex debate in terms of two sides, is to face two challenges.

The first is to assess what various parties believe to be the 'facts' about the various topics of importance in the field, such as the requirement for secure energy supplies, the management and destination of radioactive waste, climate change, the relative and absolute safety of different energy sources, nuclear weapons proliferation, the importance of research and development, and others. Many of these areas do not yield 'facts' very readily, as they involve making projections of a highly uncertain future, often on the basis of nothing more persuasive than hope, or extrapolations of a past which is itself open to different interpretations. For example, much work has been done on projecting world energy demands over the next few decades. Even so, very little can be said with any certainty, except perhaps that increases in energy demand can be expected on a global scale. Notwithstanding statements made by nuclear protagonists that nuclear power will be 'needed', or by opponents that it will not, one cannot be certain today whether a major nuclear industry, or indeed any nuclear industry, will (or should) be making a contribution to solving global energy and environmental problems in, say, fifty years' time. It is highly likely that different strategies, perhaps radically different strategies, will be followed in different regions of the world, or even in different countries within the same region, depending on local circumstances and values or traditions.

The second challenge involves a dispute between different value systems, in which the protagonists on each side hold world views which are not amenable to challenge through reference to facts or reality. Indeed, the very same 'facts', even when broadly accepted by both sides, can result in very different interpretations.

One aspect of this is the perception of different models of political structures. It can be argued that while technologies such as renewables have a tendency to decentralize control, including political control, nuclear power is inherently politically centralizing. Given the nature of the materials involved and the necessary size of the facilities, considerable state involvement may be required to ensure safety and security. The extent to which centralization or decentralization of governmental institutions is desirable is a matter of personal philosophy.

Wherever the same 'facts' lead to diametrically opposed conclusions, it is at least possible that their interpretation is driven by pre-existing attitudes to nuclear power. If this is indeed the case, then reducing the gap between the two sides may be a difficult task indeed.

This being said, decisions taken today may have a profound influence on the ability of future generations to deal with whatever reality they face. Some degree of wide-ranging forward thinking is therefore essential. While prudence might demand considerable reliance on proven and available technologies or evolutionary developments, one should not preclude the possibility of break-throughs. Indeed, careful forward thinking could pin-point where breakthroughs might be particularly useful.

## 1.3.2 The nuclear industry

One other issue needs to be addressed at this stage. Throughout the paper the term 'nuclear industry' will frequently be used. In many areas of the developed world in the 1960s, say, the phrase 'nuclear industry' could be used to describe a body of organizations and individuals who had some interest in the continued development of nuclear technology. It included govern-ment organizations researching and developing plants, the companies that constructed them, the utilities (or at least their nuclear divisions) that operated them and the companies responsible for the fuel cycle. In some countries it was closely associated with the military uses of nuclear technology. Such nuclear industries often had powerful lobbying links with national governments, which often regarded nuclear power as a symbol of technical and industrial maturity.

While such a picture can still be recognized to a greater or lesser extent in some countries, in others it has changed considerably. In some developed countries the network of companies involved in building nuclear power stations, for example, has largely broken up as new orders have stopped. (In others such networks still exist, although there has been considerable consolidation in recent years.) The research base, both in specialist agencies and in the universities, is much reduced for similar reasons. In liberalized electricity markets the companies operating nuclear stations have no particular vested interest to build more. Such companies tend to reposition themselves as electricity generators rather than nuclear power companies, and will build the facilities that are most likely to contribute to profits for their shareholders. In many countries there is little evidence of a powerful pro-nuclear lobby in the government; some indeed have taken steps to phase out nuclear power. Insofar as the term 'nuclear industry' is useful in these countries, it perhaps applies to those firms involved in the fuel cycle – manufacture, reprocessing and waste management – and the major design, construction and decommissioning companies.

This is an important point. When it is claimed that 'the nuclear industry' should take some action to develop a new plant concept or waste management strategy, for example, it is often not clear in whose interest such a development would be, and therefore who should undertake it. A new plant design, if competitive against other fuels, could be a lucrative prospect for a plant vendor. However, in competitive markets electricity generators will focus on high returns to investors, and will be unlikely to put up large sums to aid the development of such new designs. Research institutes by their nature tend not to have large funds of their own to use as they see fit. As far as the governments of at least some developed countries are concerned, perceived problems over public attitudes make nuclear investment unattractive.

As in other aspects of nuclear power, there are considerable differences between regions and between countries over the extent to which coherent nuclear industries exist. The above complications should be borne in mind when reference is made in the text to the 'nuclear industry', as it has proved a very difficult task to use the phrase consistently.

### 1.3.3 The main issues

In this paper the main issues are organized into three groups. The first group consists of relevant issues which are largely outside the control of the nuclear industry, but which serve as the 'environment' in which it must operate. This group includes:

- energy demand and choices;
- climate change;
- health effects of low levels of radiation;
- regulatory issues.

The second group consists of issues over which the nuclear industry has major influence. These include:

- the relative economics of nuclear power;
- nuclear research and development;
- safety of nuclear facilities;
- infrastructure necessary to preserve the nuclear option;
- skills requirements in the case of nuclear phase-out.

The third group consists of issues over which the nuclear industry has some but not complete control. It includes:

- nuclear proliferation;
- public perceptions/politics;
- waste management, including aspects involving politics and public perception;
- reprocessing, also including politics and public perception.

Of course, these categories do not have sharply defined edges and their effects can interact. The availability of nuclear power may have an effect on overall energy demand, while the costs of nuclear power will certainly be affected by the stance taken by regulators. Nonetheless, if only as an organizational tool, these categories are helpful.

The focus of this project will be on identifying the main areas of dispute and clarifying the issues. It will not attempt to draw firm conclusions about relevant issues such as future energy demands and how these may be met, or to say whether nuclear power will or will not, should or should not, undergo a major upturn in fortunes over the twenty-first century. It will seek to provide information rather than answers. In view of such uncertainties, one focus of this study will be on 'keeping the option open'. It is almost a truism to say that the nuclear option should be kept open in case future generations favour nuclear expansion. It is certainly hard to argue either that this generation should, or even that this generation could, entirely prevent the people of, say, 2050 building a major nuclear programme.

However, the oft-used phrase 'keeping the nuclear option open' is a difficult one to quantify. First, its meaning is not clear. To some, it seems to mean early orders of new nuclear plant in advance of market demand, in order to safeguard the skills necessary for such projects. By definition, this would require considerable action by governments, either in providing the capital for such investment or in ensuring the existence of a guaranteed market for the output of such nuclear stations, probably at guaranteed prices. To others, the nuclear option will never close. The existence of records of nuclear science means that a restart using current technology will always be possible, even if this nuclear technology should become moribund for some years.

### 1.3.4 The uncertain future

The question then arises, 'open to do what?' There are certain sets of external conditions which may appear so unlikely that the costs which would be necessary to keep the nuclear option open in a form that could respond to them would be too great; and others in which current nuclear technology would be in such demand that no further government support would be necessary. Between these extremes lies a wide range of possible futures.

As discussed in the section on R&D (3.4), the requirements of a nuclear industry of approximately the current size, based largely on replacement of existing capacity or perhaps on slow growth, might be radically different from the requirements of a nuclear industry perhaps ten times the current size. It may be fruitful, then, to consider three simple scenarios for the future of nuclear power in, say, the year 2050.

| | |
|---|---|
| 'Red' | Slow withdrawal from nuclear power as existing plants reach the end of their lifetimes in the developed world and developing countries find alternatives to currently planned nuclear expansion, leading to effectively zero capacity in the second half of the twenty-first century. |
| 'Amber' | Continuation of the present situation, including replacement of existing reactors and some new capacity in developing regions, resulting in modest growth of capacity towards 600 GW, representing some 2–5 per cent of global primary energy demand. |
| 'Green' | A major expansion to some ten times current capacity in the second half of the century, representing perhaps 15–30 per cent of global primary energy demand. |

In the Red scenario 'keeping the option open' involves no more than the preservation of sufficient skills to deal with the legacy of nuclear waste, facilities to be decommissioned, decontamination of land, etc.

In the Amber scenario a continuation of present policy might be sufficient, with new reactor designs being based largely on existing concepts. It is at least possible that there would be sufficient uranium to sustain such a scenario for some time without a requirement for the reprocessing of spent fuel. However, even here it is likely that a considerable amount of preparation, including R&D, would be necessary. Key requirements might include:

13

- development of a wider range of reactor designs, including smaller plants with lower capital outlay and shorter construction phases which are more suited to competitive electricity supply markets and possibly some developing countries, as well as large plants for more centralized systems;

- significant progress on management of waste;

- more sophisticated methods of decision-making that can engage and involve local communities and other interest groups.

To be ready to respond to the Green scenario it is likely that considerable effort would have to be expended in the near future. A major nuclear expansion might, for example, require, in addition to likely needs for the Amber scenario, any or all of the following:

- new approaches to reprocessing;

- plutonium-powered reactors;

- plutonium-producing reactors;

- partition and transmutation as an approach to managing spent fuel;

- methods of extracting uranium from seawater;

- the use of thorium.

These issues are considered in more detail in the section on R&D (3.4).

As noted above, a cursory glance at the present status of nuclear power (see Appendix 2) yields a very complex picture. Regional and national differences in policy towards nuclear energy are significant, and defy simple classification into, for example, the 'developed' versus the 'developing' world. Neighbouring countries such as Germany and France can have dramatically different political attitudes to nuclear power. Some developing countries, experiencing enormous increases in demand for energy, take pro-nuclear political stances – India and China are obvious examples – while nuclear power is making no headway in the continents of Africa or South America (with only three currently minor exceptions).

It follows, then, that the future of nuclear power, like its past and present, may be significantly different in different parts of the world. One can imagine some countries hastening the decline of nuclear generation, as Italy has already done and as Sweden and Germany have announced they intend to do (Red). Others may run existing plants for as long as it is deemed safe and economic to do so, and perhaps replace some or all of this capacity without seeing a major increase in total generation (Amber). Still others may develop their nuclear industries rapidly, alongside other energy technologies, to satisfy growing demand (Green).

Although a wide range of issues is considered in this paper, all these issues should be regarded as being in some way subsidiary to one or both of the two questions: 'what would be required to keep the nuclear option?' and 'what would the nuclear option be required to achieve?'

## 1.4 The position paper

This paper is divided into three main chapters, which correspond to the three categories outlined in section

1.3.3 above. Each chapter will be subdivided into a number of sections in which an overview of the key points is given, followed by a summary of the pro- and anti-nuclear arguments. As far as possible, these will be presented side by side and point by point, although there will be some views held by one side which do not have a directly corresponding point on the other. These arguments should be regarded as the 'boundary' views in the debate, with the vast range of serious opinion lying between the extremes. In order to avoid any accusations of bias, the order of presenting the anti- and pro- views will alternate from section to section. A list of questions for possible future consideration is given at the end of each discussion of the issues.

This approach has its limitations. Some issues give rise to a considerable range of views from within the broadly 'pro-' or 'anti-' nuclear camps. But there are some, of which the necessity of combating climate change is perhaps the most obvious, where there is considerable agreement between large sections of the pro- and anti-nuclear communities, although some other agencies are less convinced. Similarly, natural gas is the main competitor both to nuclear power and to the renewables advocated by many who oppose it.

## 1.5 Alternative approaches

The structure chosen for this paper is only one of a number of valid possibilities which recommended themselves. It would have been attractive in many ways, for example, to divide the issues covered into those which lie at least partially within the sphere of influence of governments – regulation (e.g. of health matters), safety, structure of the electricity supply industry, climate change policy, R&D, infrastructure – and those which lie more within the sphere of influence of the industry and financial institutions – economics, choice between different fuels, actual strategies of waste management, reactor size, etc. This structure would have had some similarities with the one chosen, but would, for example, have placed the focus more on governments as an audience for the findings.

Yet another approach, which seemed to the authors to have considerable merit, would have been to attempt to identify the 'boundary conditions' under which nuclear power might be expected to make a significant contribution to global energy requirements in, say, the years 2030–50. Though perhaps a little more nuclear-centred than the chosen structure, this treatment would have had many features in common with the present one. At relevant points in the paper, therefore, the question of such 'boundary conditions' will be addressed, explicitly or implicitly.

# 2 ISSUES OVER WHICH THE NUCLEAR INDUSTRY HAS RELATIVELY LITTLE INFLUENCE

## 2.1 Introduction

The future of any particular industry is not determined entirely by factors within its own control. Nuclear power is no exception. There are many issues over which the industry has relatively little influence that will affect its attractiveness to decision-makers of the future. Some of these issues concern the general requirements of an energy supply system (and in particular an electricity supply system); others involve matters of relatively pure science; still others relate to the competitor technologies to nuclear power. Resolution of these issues will depend on how a particular society chooses to meet its demands for growing energy supplies, its perceptions of clean and safe living and how to achieve it, and what precautions are to be taken against environmental degradation and especially climate change.

The project's scenario-based approach seeks not to make detailed forecasts of the future, but instead to describe actions necessary to create a nuclear industry which could respond to the wide range of possible futures, and in particular the three simple scenarios – Red, Amber and Green – outlined in section 1.3.4. In setting the bounds for likely possible futures this paper will make use of the recent World Energy Council/ International Institute for Applied Systems Analysis (WEC/IIASA, 1998) scenarios (see Appendix 1).

## 2.2 Energy choices and the 'need' for nuclear power

The topic of energy choices is inextricably linked with that of energy strategy (society's requirements from energy systems) and energy policy (the measures needed to achieve that strategy in particular circumstances). Society has perhaps five major requirements of its energy production systems:

- secure supplies;

- economic supplies;

- environmentally acceptable supplies;

- conservation of non-replaceable natural resources which have other potential uses;

- socially acceptable supplies (in terms of safety, public perception, effect on particular communities, etc.).

While all of the above are desirable in isolation, they are often in conflict. To maintain security of supply or protect regional employment, say by promoting an indigenous coal industry, may be both expensive and environmentally damaging. At least in the absence of any instruments to ensure that the 'polluter pays' for the damage done, say, by greenhouse gas emissions, the most economic sources of energy in any particular region or country may be among the most polluting, may have to be imported, and/or may have implications for local communities. Use of fossil fuels, while contributing to security of supply and low prices in many areas, results in depletion of materials which have vital roles in the production of chemicals and in transportation. Certain sources of energy which have other advantages may meet public opposition.

A nation's energy strategy must seek to strike a balance between these various requirements. It is no surprise, then, to find within energy policies variations both over time and by geographical locality as circumstances vary, and also apparent contradictions as the various contradictory requirements are addressed. Nor is it a surprise to find different fuels being used alongside each other, as part of an attempt to balance the various needs of energy strategy. Although the various fuels are often seen as being in competition with each other, this is to be tempered with a recognition that some fuels are more suited to baseload operation (by virtue of being reliable and relatively low-cost in operation, including nuclear power), others to meeting peak demand (through low capital costs – combined cycle gas turbines (CCGT)), others still to saving fuel when they are available but not to providing baseload owing to the intermittent nature of the energy source (wind, tidal, solar). The benefits of building a system which balances the advantages and disadvantages of various fuel options are clear.

In a country such as the United States current energy policy is largely driven by market forces, as economic production is probably at the top of the priority list for energy production. In France, Japan, India and China, where for different reasons there are concerns over the security of future energy supplies (whether because of a shortage of indigenous resources, limitations on transportation infrastructure for moving coal to centres of economic activity, significant increases in energy demand or some combination of these), providing reliable energy for the future tends to take precedence, sometimes eschewing the cheapest source of energy available to some extent. Norway, among other countries, has introduced a significant carbon tax which serves as an incentive to reduce greenhouse gas emissions, though this inevitably adds to the energy costs faced by domestic industry. Trade union activity and considerations about the social effects of rapid closure of coalmines on local communities have been factors in determining coal policy in Australia, the UK, Spain and Germany.

Further, the 'need' for any particular source of energy is a problematic concept, at least in a literal sense. The real need is for secure, economic, environmentally acceptable supplies of electricity on a large scale. The nuclear industry has claimed from time to time that nuclear power is going to be needed, to meet the world's energy demands, to act as a hedge against overdependence on imports of hydrocarbon fuels, or to safeguard the atmosphere from unmanageable releases of greenhouse gases. However, it is patently untrue to say that the future of the human species would be impossible without an ongoing nuclear generation industry (or any other individual technology, for that matter). In a softer sense, to claim there is a 'need' for nuclear power is to assume that other alternatives will not be sufficient to meet future energy demands without unacceptable cost and environmental impacts.

In reality, the debate centres on whether, on balance, society in the future would be 'better' or 'worse' if there were a nuclear industry of a particular size, and thence what the range of sizes of a nuclear industry should be (a range which might, of course, include 'zero'). One can imagine circumstances – rapid development of competitive solar power, or of economic methods of scrubbing carbon dioxide from the waste streams of fossil-fired electricity plants, for example – in which nuclear technology might look more unattractive than today, at least assuming there are no matching innovations in the nuclear field. Alternatively, if methods of exploiting renewables which are more attractive (politically, technically and/or economically) are not forthcoming, and the effects of climate change prove to be in the upper regions of current estimates, nuclear power may look highly attractive. Clearly, success or otherwise of research to contain radioactive waste, for example, or to reduce the capital investment necessary for nuclear technology, would affect the relative attractiveness of nuclear investment. But would this make for a 'better world'?

This is an enormously difficult question to approach, for at least two distinct reasons. First, there are as many different conceptions of what would make a 'better world' as there are commentators on the issue. For some, a high-tech world of free enterprise would do most to liberate people; to others a much simpler, low-technology lifestyle would serve to re-establish the interpersonal relations at the centre of the human experience.

Second, the area is one of enormous uncertainty, both about current scientific issues such as the health effects of low levels of radiation, and about the future, e.g. demand for energy, severity of climate change, options for sequestering carbon dioxide from fossil fuel exhaust streams, development of renewables, etc.

The issue, then, becomes one of providing a sequence of options that might be applied, with varying degrees of success, to a range of possible futures. Issues of this nature have led to the growth of scenario planning as an approach to longer-term policy matters, including investment.

The key aim is to reduce the risk that humankind may be unable to respond appropriately to future conditions. At one extreme is the risk that suitable technology might not be available to provide for world energy demands, with possible political destabilization or environmental crises. At the other is the risk that unnecessary levels of resources might be wasted in creating technologies which will not be required or which might entail unnecessary environmental costs. Finding a middle path between such extremes when the future is so uncertain is the challenge to planners and decision-makers.

It is difficult to argue against 'keeping the nuclear option (or any other option) open' for possible future use as circumstances dictate. In effect, 'keeping the option open' is an insurance policy offering more flexibility in dealing with future, unpredictable changes. Flexibility is expensive. If, say, the developing world were to face severe difficulties in satisfying growing energy demand, with the potential political and economic instability such shortages would entail, or if climate change proved unmanageable, then the costs of keeping a nuclear insurance policy might be judged money well spent. On the other hand, if the costs of keeping the nuclear option open were seen as excessive and the number of scenarios in which the option would be required were limited, the 'value for money' of the nuclear insurance policy (or some forms of it) might appear too poor.

It has been argued that nuclear power, in its present form, is a relatively inflexible technology requiring large infrastructures, long timescales and significant and largely irreversible resource commitments. By contrast, renewables can be built in small units, while natural gas has much lower capital costs per unit of electrical output. A more flexible nuclear industry, with smaller units, lower capital costs and shorter construction periods, may be more appropriate for some models of liberalized markets. Against this, the economies of scale enjoyed in some countries such as France and Japan may make large plants relatively attractive, given the arrangement of the electricity supply markets in those countries. Large plants may similarly remain appealing in countries where electricity demand is growing rapidly.

Projections of energy use are dependent on assumptions about economic growth, energy intensity of economies, etc., which are by their nature dependent on a variety of factors, known and unknown. When projections are made of associated issues, such as the fuel mix or emissions of carbon dioxide and other environmental emissions, uncertainties can be even greater. The history of projections of energy demand is full of examples of both overestimates (particularly in the 1970s) and, perhaps less commonly, underestimates. The speed with which projections can change is enormous, and considerable caution must be applied to any future energy projections.

Additional levels of complexity are introduced by the fact that there is a wide range of different approaches discernible even within the developed world. It is likely that radically different policies may be pursued by even neighbouring countries when addressing similar concerns about security of energy supplies, costs, environmental protection, etc. Similarly, the timescales on which governments and industry work seem to vary significantly, some Asian countries working to very long timescales, others in the market economies working to the end of the next electricity contract (or sometimes to the next general election).

A further aspect of this issue is the particular role as a provider of baseload electricity which nuclear power plays in many developed economies. It would appear unlikely that renewable sources such as wind or solar power could easily take on this role, as they are reliant on intermittent sources of energy, although other renewables, such as biomass, are not. At the same time, however, it is possible that the need for large baseload capacity might decrease in some developed countries as smaller local generating plants gain importance in liberalized markets.

## 2.2.1 Energy futures

Policy planners have increasingly addressed this problem by exploring alternative futures through use of scenarios. Scenarios are alternative views of the future, which should provide a 'feel' for likely levels of uncertainty. Each scenario has to embody consistent assumptions. If the assumptions in different scenarios differ widely but are still within the realms of plausibility, then a broad range of possible outcomes emerges.

One example is the scenario approach used in *Global Energy Perspectives* (WEC/IIASA, 1998 – see Appendix 1). This involves three families of scenarios and a total of six variants. On the assumption that world population doubles from 1990 levels by the middle years of the twenty-first century, the scenarios span the ranges shown in Table 1.

**Table 1: Range of estimates (1990 = 1)**

| Year | 2050 | 2100 |
|---|---|---|
| Economic growth | 3–5 | 10–15 |
| Primary energy demand | 1.5–3 | 2–5 |
| Nuclear power capacity | 1–6 | 0–19 |

These ranges assume that world primary energy use (including non-traded energy) for 2050 will be between 14 and 25 billion tonnes oil equivalent (btoe), and for 2100 between 18 and 45 btoe, compared to 9 btoe in 1990. (See Appendix 1 for more details.) If such amounts of energy are to be produced, several challenges will have to be overcome. In some countries, a continuation of the trend towards gas could lead to concerns about overdependence on imported energy. Even if coal continued to lose market share, persistent large-scale use would have considerable environmental implications, unless some efficient method of sequestering carbon dioxide emissions could be developed. By contrast, although considerable expansion in nuclear generation would lead to reduced emissions of greenhouse gases, it would, *inter alia*, increase the need for waste disposal, and might have proliferation implications.

## 2.2.2 Energy security

Energy is at least nominally treated simply as a traded commodity by several governments, and such a view has been gaining in prevalence in many developed countries since around 1990. However, there are two key ways in which electricity in particular differs from many marketable goods.

First, given current technology, electricity cannot be stored in large quantities. In some countries trials have been carried out of techniques such as pumped storage (use of off-peak electricity to pump water uphill for availability as hydropower at times of high demand) but they have generally proved to be rather expensive. More esoteric approaches, such as use of off-peak or specially generated electricity to electrolyse water, with the resultant hydrogen being used in small-scale power or transport applications, or using superconductors as a method of storing electricity, are at present at a relatively early stage of evaluation.

Second, interruptions in electricity supply are likely to have far higher economic and other consequences for the consumer than for the producer. In many markets failure to satisfy demand causes roughly equal negative consequences (disutility) both for producers, who lose out on potential profits and perhaps goodwill, and for the consumers, who are not able to enjoy a product for which they were prepared to pay. However, loss of electricity supply (the same can also apply to other forms of energy to greater or lesser extents) can be extremely damaging to consumers, depending on the length of disruption to supply. The loss, say, of a freezerful of food, or a day's production at the factory, or energy to the operating theatre far outweighs the price of the relevant units of electricity.

It has therefore been a concern of governments, from time to time and place to place, to play a role in ensuring secure long-term supplies of energy. For a variety of reasons, this may particularly be the case when energy security looks under threat. Since 1990 the governments of some developed countries have to a greater or lesser extent stepped aside from ensuring secure energy supplies. However, over the last century, and indeed in much of the world today, it has been more usual for governments to play a much more central role in provision of energy supplies, and especially of electricity.

There are a number of possible threats to secure supplies of energy in general and electricity in particular, of which perhaps three can be taken as particularly important.

Some countries, though not facing significant increases in the demand for energy, are nonetheless not rich in natural and economically attractive energy resources such as oil, coal or gas. Japan and France fit into this category.

Other countries are facing major increases in demand for energy, and although they may have indigenous energy reserves, especially coal, these are sometimes considerable distances from areas of major economic activity. India and China are in this position. A major expansion of coal-based energy generation in these countries would require enormous investment in transportation and transmission infrastructures. Similarly, in Russia the oilfields and gasfields tend to be in the east of the country and the economic centres in the west; the total length of main oil and gas pipelines in Russia is close to the distance between the earth and the moon (Gagarinski, 2000a). The costs involved in transportation of coal, oil or gas, or transmission of electricity over such large distances, clearly improve the relative economics of nuclear power, where transportation of fuel is a much smaller factor.

It may only be those countries with stable energy demands and with significant fossil reserves of their own or easy access to apparently reliable imports that can afford to place security lower down the priority list for their energy strategies, and therefore adopt policies in which the market plays more of a part. The UK was at the forefront in liberalizing the energy market and a number of other developed countries have followed.

The advent of natural gas (see section 3.3) as a major energy resource has been a significant factor in developing perceptions that energy supplies, at least in the medium term, are unlikely to be subject to dislocations such as those which were observed in the 1970s. However, this is not to say that security of supply will never be a preoccupation in the developed world again. One possible scenario, for example, might combine:

- failure to develop renewables, perhaps because of technical limitations;
- low levels of investment in new nuclear stations in the medium term;
- continued global economic growth fuelled predominantly by oil and gas, perhaps because current projections of global warming prove to be exaggerations.

Pressure on world oil and gas supplies would be significant, and the risk of shortages leading to economic collapse, cross-border wars for resources, etc. might well be considerable.

Another scenario in which governments might choose to become more directly involved in energy provision could involve escalation of the direct action against fuel duties which spread through Europe in mid-2000 and caused significant economic dislocation within a very short space of time.

It has also been noted that use of natural gas, a versatile resource, for electricity production incurs high opportunity costs; some commentators question the ethics of using such a resource if others are available and suitable.

Further, in the very countries that have liberalized their electricity markets because perceived external threats to security are small, a third threat to secure supplies of electricity may be emerging, or rather re-emerging. In liberalized markets it is practically impossible to impose a duty to supply on any player in the same way as can be done in monopoly-based systems. In most developed countries grid systems were designed for, or at least evolved alongside, centralized large-scale generating stations.

Liberalization is leading to investment in large numbers of relatively small-scale generating units. These units may be aimed at the power requirements of a small area, with local customers requiring import of electricity from a grid only at times of high demand. A significant proportion of the new generating plants will be powered by renewables, often with intermittent power production. It is not clear, and it may even be unlikely, that secure electricity supplies can be guaranteed in these circumstances.

For some major industrial users of electricity, running a risk of less secure supplies in return for lower prices may be an attractive option. It is more difficult to determine where that balance might lie for small-scale consumers of electricity. It is at least feasible that prolonged power cuts would lead to unacceptable political consequences for the government of the day, and that a reversal of market liberalization might result. (In effect these governments might 'renationalize' the duty to ensure secure electricity supplies.)

For these reasons, perspectives on the security of energy supplies differ from region to region, and indeed from country to country, and this affects energy choices.

### 2.2.3 Renewables

Nuclear power is not the only carbon-free energy technology. The range of possible commercial technologies under the general heading 'renewables' includes hydropower, by far the most widely exploited at present. Among the others, the most promising at present appear to be biomass ('carbon-neutral' rather than carbon-free), solar and wind power. Further possibilities include tidal and wave power, geothermal energy, waste burning or use of methane from landfill. Though these are often referred to as the 'new renewables', in fact only photovoltaics is truly new, many of the others having been used in some form for many centuries.

The theoretical potential for renewables is enormous, far higher than feasible world energy use. As is the case for all sources of energy, however, the difficulties lie in the conversion of energy potential into useful forms. Renewables are generally held to share two of the advantages of nuclear power in that they do not depend on using limited fuel supplies which have other beneficial uses, and in operation they do not contribute to the great atmospheric pollution problems, notably climate change.

Comparisons of the overall safety of renewables and nuclear power are difficult to make. Some renewables are associated with a significant number of small accidents in the construction, installation and exploitation phases (and the possibility of enormous catastrophe in the case of hydropower, as well as major environmental and population disruption), while nuclear power is very safe in normal operation but contains the potential for very major accidents.

These may be counted as advantages for renewables over nuclear technology in at least some circumstances. Renewables do not result in production of radioactive waste or releases of radioactive discharges, and do not have implications for nuclear weapons proliferation. Further, they can be built in relatively small, modular units, unlike nuclear plants which are always likely to need to be of a certain minimum size. The liberalization of electricity supply markets in many developed countries is leading to a growth in demand for small generating units, in preference to the 1,000MW-plus units which were standard in most developed countries in the 1970s and 1980s. This trend is clearly favourable to some renewable sources of energy.

Further, their advocates hold that many renewable technologies, being less technically demanding than nuclear power, are inherently more attractive. Indeed they query why the nuclear option should be pursued at all when more technically benign options are available.

However, there are also questions over renewable technologies, stemming largely from the very low power density and intermittent nature of the fuel sources. These include:

- large land or sea/lake requirements to collect significant amounts of energy, resulting in local environmental damage (it is estimated, for example, that a 1 GW(e) power station – a power station with electrical output of 1,000 MW – would require 20 km$^2$ of solar cells, a wind farm covering more than 50 km$^2$, or 4,000 km$^2$ of forestation for biomass) and potential local protest;

- associated questions about availability of land, which is often required for other potential uses such as agriculture;

- considerable capital investment per unit of installed capacity (though low operating costs), making renewables relatively economically unattractive in competitive markets when compared to sources with low initial costs;

- material and energy requirements in the construction phase, resulting in indirect emissions of greenhouse gases and high costs of energy production (Uchiyama, 1995) – this point is also relevant to nuclear power;

- the intermittent nature of many renewables, implying the need to provide back-up capacity (fossil, nuclear or major overcapacity of renewables) for occasions when wind, tidal, solar or similar energy might not be available, with obvious cost and environmental implications; also the difficulty of integrating intermittent sources of energy into a grid in the absence of a large-scale method of storing electricity;

- the apparent inability to support major transport applications, e.g. shipping;

- questions over how appropriate renewables will be for the increasing numbers of the world's citizens living in the new mega-cities, where a high proportion of energy, especially electricity, must be generated outside and imported.

There is a wide diversity of views about the potential for renewables in the medium term. These range from commentators who are certain that large-scale renewable programmes will be economic and acceptable in a very short time, to those who are sceptical as to whether renewables could ever significantly supplant current sources of electricity, notably natural gas and nuclear power, especially for providing electricity to large and densely populated cities.

The WEC/IIASI scenarios all see a steep increase in renewable generation for tradable energy – see Appendix 1. However, the most optimistic scenarios have renewables producing 39 per cent of primary traded energy by 2050. In all the scenarios the total amount of energy to be produced by non-renewable sources will be considerably higher than in 1990.

### 2.2.4 Energy choices: the pro-nuclear and anti-nuclear positions

| PRO | ANTI |
|---|---|
| All credible projections suggest that energy demand will continue to increase significantly into the foreseeable future. World population is growing rapidly, and many developing countries aspire to and expect to enjoy Western standards of living. | Demand-side management, coupled possibly with a simpler lifestyle, could reduce demand for energy very considerably over current projections. |
| Although fossil fuel reserves are likely to be considerably greater than now proven, it is unlikely that fossil fuels alone will be able to meet growing energy demand. Furthermore, the environmental effects of a major increase in fossil fuel use, and in particular the | The environmental effects of a major increase in fossil fuel use, and in particular the increase in emissions of greenhouse gases and damage caused during extraction, preclude such increases. |

consequent increase in emissions of greenhouse gases, preclude this course.

Improving energy efficiency will not involve a proportionate reduction in energy use, as much of the benefit may be taken in increased economic activity.

Energy efficiency measures, particularly if combined with higher energy prices, could reduce projected demands significantly.

Electricity use is increasing as a proportion of overall energy use in all regions.

Electricity use is especially amenable to improvements in efficiency in both production and end use.

Renewable forms of energy, though promising in some cases, cannot take over the role of fossil fuels, particularly in the short and medium term.

The potential for renewable energy has consistently been underestimated, and may be very considerable. Further, renewable technologies are of their nature more technically benign and less demanding.

It is essential to retain a nuclear infrastructure – educational base, research network, heavy engineering against a possible resurgence in the demand for nuclear power. State support might be required in some form.

The costs of retaining a major nuclear infrastructure ahead of demand which may never materialize would be too high to be supported by governments. Private industry, recognizing the absence of an economic case for nuclear power, will not support such an infrastructure.

It may be important to increase the diversity of available nuclear power

Most new electricity demand will be in developing countries, where use of plant designs, in terms e.g. of unit size. New designs of nuclear stations, especially involving lower capital costs per unit, may be appropriate for use in some countries, while large plant designs remain appropriate in others. All non-fossil forms of electricity generation are capital-intensive.

nuclear power is unsuitable owing to its high capital costs, requirements for large-scale grid networks and technological challenges.

### 2.2.5 Questions for further consideration

- Under what circumstances would concerns about security of supply re-emerge in those developed countries which have liberalized their electricity markets?

- What responses might governments make in these circumstances?

- What is the realistic potential of renewables in the medium to long term?

## 2.3 Global climate change

There now seems to be considerable 'qualitative certainty' about climate change. It is believed by many in the academic community that some level of anthropogenic climate change is highly likely and may be detectable already, although there remains a wide range of estimates as to likely severity. Many governments have also accepted the argument, as demonstrated in international agreements such as the Kyoto Protocol.

Ironically, the threat posed by global climate change is taken at least as seriously by the anti-nuclear movement as by nuclear power's supporters. By contrast, there are some, neither pro-nuclear nor anti-nuclear, who have argued against immediate major steps being taken to reduce the risk, e.g. some major industrial companies, especially in the United States. Advocates of sequestration of carbon dioxide argue that the environmental effects of fossils fuels can be reduced significantly.

It will be some time before detailed projections of the effects of global warming and climate change can be offered with any confidence. However, the range of present projections includes possibilities in which climate change would dwarf any other environmental problem and represent an enormous challenge to ecosystems, water supplies, coastal communities, insurance companies, etc.

Climate change is believed to be associated with a wide range of emissions, of which carbon dioxide, methane and nitrous oxide are the main ones associated with energy. Carbon dioxide is emitted whenever fossil fuels are burned, with the ratio of emissions from electricity production from coal, oil and gas being roughly 6:5:3. Methane is released because of gas production, coalmining and leaks in natural gas pipelines, as well as from other non-energy sources such as agriculture and domestic waste. Nitrous oxide (a far less important greenhouse gas than carbon dioxide or methane) is a minor product of coal burning.

Reducing greenhouse forcing (the tendency of greenhouse gases to cause global warming) will be an enormous and complex task. The Intergovernmental Panel on Climate Change (1995), supported by other bodies such as the UK Royal Commission on Environmental Pollution (2000), have calculated that to stabilize concentrations of carbon dioxide at double their pre-industrial level would require a global reduction in emissions of some 60 per cent from 1990 levels by 2050, and 80 per cent by 2100. For comparison, complete achievement of the aims of the Kyoto Protocol would result in global greenhouse gas emissions some 30 per cent higher in 2008–12 than they were in 1990.

When climate change is considered in the context of growing demand for energy, it follows that major increases in some alternatives to fossil fuels – demand reduction, renewables and/or nuclear power – will be necessary unless some method of reducing greenhouse gas emissions from fossil fuels becomes viable.

The development of methods of sequestering carbon dioxide emissions from fossil-fuel-fired power stations might significantly increase the continuing attractiveness of fossil fuels for power production. Sequestration is the technique whereby emissions of carbon dioxide associated with use of fossil fuels are removed. This may be done either indirectly, by removing carbon dioxide from the atmosphere by vegetation, especially new forests; or directly, by capturing the carbon dioxide at source and transporting it to some site for disposal. Sites under consideration include:

- former oilfields, gasfields or coal mines;
- geological formations such as deep aquifers;
- deep oceans.

A number of technical issues remain to be resolved. Significant leaks during transportation of carbon dioxide from source to sink could be dangerous. Ocean disposal could cause localized environmental problems, while the integrity of sinks would have to be established. Further, at present such methods are expensive and would add between 40 per cent and 100 per cent to the costs of power production (Audus and Freund, 1997). The most expensive step, representing some 80–90 per cent of total costs, is the separation and capture of carbon dioxide from waste streams.

That said, the capacity of geological sinks for carbon dioxide storage appears enormous, representing some centuries of global carbon dioxide emissions at present rates. Improved techniques could well be developed which would reduce the costs of separation and capture considerably, but this cannot be assumed. In the absence of more cost-effective and proven sequestration technologies, it is calculated that at least half of energy generated in 2050 will have to come from carbon-free technologies.

Conversion from coal to natural gas allowed a number of countries, notably the United Kingdom, to achieve relative reductions in carbon dioxide emissions during the 1990s. However, such reductions are likely to prove short-term, and emissions will probably begin to rise again once all coal has been replaced (assuming electricity demand increases, as is expected even if overall energy use remains broadly constant). In the longer term and in the absence of cost-effective sequestration options, reductions in emissions are likely to be achieved by reductions in energy demand, and/or by switching to zero-carbon fuels such as nuclear power and renewables.

Considerable improvements in energy efficiency, both in production and in end use, seem possible. However, even assuming such improvements were made, their likely effect is by no means clear, and it cannot be assumed that the result would be significant reductions in energy use. Historical evidence strongly suggests that at least some of the benefits of such improvements might be taken in increased economic activity rather than in reduced energy consumption. Increased energy prices would in all probability lead to reductions in energy intensity of economic output, especially if coupled with tighter regulations (as happened, for example, in the United States in the 1970s – PCAST, 1997), but the extent is difficult to estimate with any confidence.

Even if fuel switching proved the most appropriate response, a wide range of nuclear and renewable technologies could contribute. The precise pattern of responses will vary from country to country, and will depend in part on the relative success of research, development and deployment programmes for the various technologies. One way of encouraging fuel switching would be to introduce market instruments to internalize the costs of emitting greenhouse gases. The two main candidates are carbon taxes and tradable carbon emission permits. The former have the advantage of setting a clear cost, which will aid businesses in their planning to reduce emissions, the disadvantage being that they cannot deliver guaranteed reductions in emissions. The latter have the advantage of certainty in emission reductions (given accurate monitoring and suitable non-compliance regimes) but do not set a firm price on greenhouse gas emissions. In principle, the price of the tradable permits would be equivalent to the marginal cost of reducing carbon dioxide emissions. This would be the lowest-cost option from among the various possibilities – renewables or nuclear power, sequestering carbon dioxide, or reducing demand. However, in limiting the total volume of greenhouse gases that could be emitted, introduction of tradable permits might be expected to create segments of the electricity supply market which must be met by non-fossil sources. This would presumably reduce the economic risk of investing in nuclear or renewable plants, both of which tend to be highly capital-intensive. It is likely, however, that the introduction of market instruments such as carbon taxes and/or tradable emission permits would be opposed by elements within the fossil fuel industries.

The 'hydrogen economy' is attracting much research and policy attention. The concept is certainly attractive. The direct fuel for transportation and for small-scale electricity production would be hydrogen used in fuel cells. The only significant downstream emission would be water. However, at present most of the world's commercial hydrogen is produced by steam-reforming natural gas, producing carbon dioxide. For the hydrogen economy to lead to major climate change benefits (unless a cost-effective method is found of removing and storing this carbon dioxide) it would be necessary to make the hydrogen in another way, perhaps by electrolysing water using electricity from purpose-built nuclear power stations or renewable plants. In any case, the infrastructure necessary to support a major use of hydrogen in these ways is likely to take considerable time and investment to install.

## 2.3.1 Climate change: the anti and pro positions

### ANTI

Climate change is a severe environmental threat, and significant action will have to be taken in the near future to enable the world to respond, should the effect prove as damaging as is currently predicted. However, other environmental problems, such as radioactive pollution, may be just as serious. Increasing emissions of radioactive materials, risk of nuclear accident, etc. as a response to fears about climate change would be unacceptable.

Energy efficiency improvements can reduce the greenhouse gas emissions problem by a very significant degree.

Nuclear power should not benefit from introduction of market instruments to reduce greenhouse gas emissions, in view of its other dangers and environmental disadvantages.

### PRO

Climate change represents the most severe environmental threat, and significant action will have to be taken in the near future to enable the world to respond, should the effect prove as damaging as is currently predicted. Though other environmental problems should not be ignored, climate change is by far the most serious; combating it should receive the highest priority.

Improvements in energy efficiency alone will not be sufficient to reduce emissions of greenhouse gases to sustainable levels, and may even have a disappointing effect on overall energy demand.

Taxes or a system of tradable emission permits should be introduced to ensure that fossil sources of energy pay for their environmental effects, and to encourage reductions in energy use or fuel switching, whichever is the more economic.

Many new renewables, notably solar power, biomass and wind, have enormous potential to reduce greenhouse emissions, if in receipt of proper research, development and deployment (R, D&D) support.

Nuclear power is the only large-scale technology, apart from hydropower, which is a proven way of reducing greenhouse gas emissions from electricity production.

Nuclear power involves production of greenhouse emissions during mining, construction, transportation, fuel manufacture, waste management, etc.

Although use of nuclear power does result in some emissions, these represent less than 3 per cent of emissions associated with the coal cycle (Rashad and Hammad, 2000). Renewables, such as photovoltaics, also generate emissions of greenhouse gases over their lifecycle, though again considerably less than fossil fuels.

Nuclear power has proved an extremely expensive way of reducing greenhouse gas emissions. A nuclear programme capable of making a major contribution to greenhouse gas emission reduction would lead to unacceptable consequences in terms of waste production, accidents and weapons proliferation.

Nuclear power can be a cost-effective way of addressing greenhouse gas emissions. Present technology has high capital costs which make it unattractive to the private market in some countries, but new designs are feasible with lower initial costs. Experience of the early 1970s shows that the world can accommodate a rapid programme of nuclear construction.

### 2.3.2 Questions for further consideration

- What are the comparative greenhouse gas implications of various nuclear, fossil and renewable energy technologies on a lifecycle basis (i.e. extraction, refining, transportation, use, waste management)?

- How would the cost-effectiveness of nuclear power – both present technologies and possible future developments – as a greenhouse gas mitigation strategy compare with other strategies that might be available in the medium to long term, i.e. demand reduction measures, renewables and sequestration of carbon dioxide?

## 2.4 Low-level radiation

The importance of the issue of the health effects of low-level radiation cannot be exaggerated. Were it not for the fear that very low levels of radiation might be associated with human health problems (particularly cancer and genetic effects) and with wider ecological damage, many of the present difficulties faced by nuclear technology would not arise. It seems certain that public fears about the implications of nuclear power (such as radioactive waste and the safety of nuclear installations) would be considerably allayed. Regulations governing routine and accidental releases of low levels of radioactive materials would be less stringent, leading to considerably reduced costs of nuclear generation, as the capital costs of construction and lead-times would be much reduced. Further, fears about radiation, often exacerbated by past actions of the industry itself, are a key factor in determining public attitudes to the technology as a whole.

Radiation is a ubiquitous feature of the natural environment. People receive 'background' doses from naturally occurring radon gas, which can accumulate in living spaces, especially if they are well insulated; from direct radiation from rocks, soil and building materials; from naturally occurring radioactive materials in the diet; and from cosmic radiation, largely but not entirely absorbed by the atmosphere. There can be enormous variations in the natural levels of radioactivity from one area to another, depending on geology, altitude, etc. Some of the beaches of southern India, for example, deliver doses more than twenty times the average background dose, yet are regarded as havens of health by local and visiting people (Sohrabi, 1998).

There is little dispute that levels of radioactivity released into the environment from nuclear installations are small compared to these background levels, though there is some dispute as to whether natural and artificial radiation have similar biological effects. Doses caused by the civil nuclear industry to most of its workers, and to practically all people living offsite, are therefore much smaller than doses received because of natural radioactivity. Further, of the doses caused by human activities, by far the largest are associated with medical uses of radioactive materials.

There is considerable certainty about the short-term effects of high doses, but a wide range of views about low-dose rates. At one end of the spectrum is a belief that radiation is more dangerous, dose for dose, at low-dose rates than at high-dose ones. Some argue for a 'linear no-threshold hypothesis', which holds that any exposure to radiation carries a proportional risk. Next comes a belief that low levels of radiation have no effect one way or the other, and finally there are those who hold that below a certain dose rate radiation becomes beneficial.

It is perhaps unlikely that statistics alone will have sufficient power to settle conclusively which of these models is the true one; the effect seems too small to be distinguished from random fluctuations in cancer rates. The linear no-threshold hypothesis has been adopted by international regulatory bodies to guide policy over radiological protection, in line with the precautionary principle that where uncertainty exists over environmental discharges it should be assumed that they are harmful until there is evidence to the contrary.

Much of the current understanding of the health effects of low levels of radiation has been developed from a long-term study of the survivors of Hiroshima and Nagasaki. The most significant dose sustained by this population came from outside the body and was instantaneous. There may be differences between such doses and longer-term doses originating from radioactive materials within the body. Both animal studies and studies on other populations suggest that different types of dose are broadly comparable, but the question is not settled.

Improved understanding of the mechanisms which cause cancer should be sought. Recent work on the biological effects of low levels of radiation has raised the possibility that the models used in predicting the health effects of radiation at low-dose rates may require revision. Very low doses may result in genetic changes which might be reflected in later generations (Wright, 1998). It is not clear at present what the outcome of such changes may be, but the possibility of affected cells becoming cancerous cannot be discounted. By contrast, low doses may keep the cell-based DNA repair mechanism in a state of readiness to combat other cancer-initiating events, and so be protective against the development of cancer. There seems to be some evidence for both of these positions.

There is increasing concern in some quarters about the potential effects of radioactive releases on the wider ecology of an area. These concerns share many features with worries about effects on human health (e.g. effects of radiation on non-human life in particular regions or circumstances), but they differ in their focus.

## 2.4.1 Radiation: the pro and anti positions

| PRO | ANTI |
| --- | --- |
| Radiation doses from the activities of the civil nuclear power industry are very much smaller than background doses, which do not seem to be associated with health problems, and doses from other human activities. | Natural radioactive materials may not be a good guide to the effects of human-made radioisotopes, which may be absorbed into the body in different ways. Such materials represent a risk not only to humans, but also to wider ecological systems. |
| Radiation at low-dose rates is unlikely to be a serious health hazard, if one at all, to humans or other forms of life. There is some evidence that very low levels of | Radiation remains a risk down to zero-dose rates. There is evidence that very low-dose rates may be more dangerous pro rata than slightly higher doses. The |

radiation may be beneficial. The assumption that radiation remains dangerous towards zero-dose rate (the 'linear no-threshold' model) has been adopted in accordance with the precautionary principle, not as a statement of physical reality. marginal benefits associated with nuclear power are overwhelmingly outweighed by the potential risks connected with exposure to low-level radiation through routine activities, accidents and waste disposal.

---

### 2.4.2 Question for further consideration

● What is the state of current thinking about the effects of low levels of radiation on human and other life, including issues such as the evidence for genetic effects and for possible protective effects of radiation at low doses?

## 2.5 Regulatory issues

There are other economic issues of key importance to capital-intensive forms of energy. The nuclear power field has been subject to regular and often significant changes in regulation, e.g. of emission limits, safety standards, etc. Indeed, it is difficult to think of another industry which responds to such a wide range of statutory and voluntary bodies, on a national and international level. These include organizations such as WANO (the World Association of Nuclear Operators), an international non-governmental grouping of companies operating nuclear plants, formed after the Chernobyl accident, which aims to share best practice and technical information between nuclear operators, and INPO (the Institute for Nuclear

Power Operators), a US-based association formed after the Three Mile Island accident. These carry out audits which, though voluntary, are often relied upon by regulators as indicators of management efficiency.

Two categories of regulations can perhaps be differentiated. The first is regulation of discharges, doses, etc., and is subject to considerable international agreement, especially through the work of the International Commission on Radiological Protection (ICRP). If the present estimates of dose–response relationships is broadly correct – and this is a topic of dispute in some circles – then permitted doses represent an extremely small health risk, much lower than that associated with many other common activities. The second is the regulation of general issues such as plant design and operation, and here significant differences are to be found in different countries. Safety regulation is also design-specific.

Radical differences between regulatory approaches for reactors, transportation flasks, etc. in different countries can be expensive, and ultimately act as barriers to trade. The adoption of the Westinghouse SNUPPS design for the pressurized water reactor for Sizewell B in the UK, for example, involved some £700 million of one-off costs associated with redesign to meet UK safety regulations (NUCG, 1994). By contrast, however, lax regulation of safety anywhere which allowed a significant accident to occur could have major economic consequences for all operators of nuclear plants.

The regulatory regime can therefore have three broad types of economic effect on a technology such as nuclear power.

- The costs associated with regulation can increase unit costs of the technology.

- The risk of future regulatory changes can add to the economic risk of a project (e.g. by a future need to retrofit expensive equipment), and hence to the rate of return demanded by investors. The simple threat of delay in construction or in awaiting an operating licence, thereby tying up a large capital investment, can be very serious.

- Lax regulation resulting in increased risk of accident in one country could threaten nuclear investments in all countries.

These observations have led to considerable attention being paid to reducing regulatory risks while maintaining regulatory effectiveness. However, although the idea of an international regulatory regime may be attractive in some ways, it also presents several less attractive features. Regulators need to have an intimate understanding of the plants for which they are responsible, so organization of regulation along national lines is likely to be more effective.

Hence the key, at least in the short to medium term, might be harmonization of national regulations rather than attempts to create an overarching and possibly ineffective international regime.

## 2.5.1 Regulatory issues: the anti and pro positions

*ANTI*

*PRO*

Regulation should be maintained at least at present levels. Ideally, discharge limits should be reduced to zero immediately; if technology capable of doing this is not available, then nuclear activities should stop.

The industry is over-regulated, to the extent of being economically disadvantaged. The same standards of waste management, routine emissions and plant safety should be applied to all power systems. Regulatory certainty rather than a constantly changing regime is essential for economic projects with long pay-back periods.

Any internationalization of regulation should harmonize to the toughest current regulations.

Harmonization of national regulatory standards, analogous to the aircraft industry, would ensure worldwide safety, while allowing for economies of scale in power plant construction.

## 2.5.2 Questions for further consideration

- What are the advantages and feasibility of introducing an international regime of regulation and inspection, and how could its credibility be assured?

- How could appropriate and fair terms of reference for such a regime be defined and established?

- Who should pay for an international system of regulation and inspection?

- Can, and should, nuclear regulations be harmonized with those of other industries? What risks, if any, are unique to nuclear power?

# 3 ISSUES OVER WHICH THE NUCLEAR INDUSTRY HAS MAJOR INFLUENCE

## 3.1 Introduction

The topics under this heading are those which might be considered while attempting to answer questions such as: 'What does the nuclear industry have to do to make itself more attractive'? It is possible to imagine a world in which a major nuclear revival occurs owing to evolutions of current nuclear technology, despite real and perceived problems. After all, nuclear power did expand rapidly through the 1970s and 1980s. Even in the 1990s it was the fastest growing of the world's major energy sources (output growing by 30 per cent – BP Amoco, 2000), although the latter phases of this expansion were fuelled more by orders made in the period up to about 1980 rather than by a large number of new orders. A return to the conditions of the 1970s – serious perceived shortages of fossil fuels, high oil prices, heavily regulated electricity supply industries – might result in a return to nuclear orders in those countries most affected.

However, even a return to the conditions of the 1970s might not lead to such a response. Fears about radioactive waste management and nuclear safety are more deeply entrenched in many countries today than they were in the mid-1970s. (Against this, fears about the effects of global warming and climate change have become much greater since about 1990). In some countries, notably the United States and the UK, the economics even of existing nuclear power plants have been disappointing up until the last few years. Installation costs have increased more rapidly than inflation since the mid-1970s, although operating economics have improved significantly over the last decade. The average construction time for nuclear power units grew from approximately 60 months in 1955–65 to between 100 and 120 months in 1995–9 (van den Durpel and Bertel, 2000). In other countries – Belgium, France, China, Japan, Sweden – economics have been more favourable. Nonetheless, electrical utilities are not choosing to build nuclear stations in economies where liberalization of the electricity supply is well advanced. Some of the renewable forms of energy are closer to large-scale commercial exploitation.

As proposed in section 1.3, a focus on three simple nuclear scenarios may be fruitful. The three scenarios have different implications for R&D and for timing of investment. It seems likely that a rather different kind of nuclear industry might be required for the scenario in which major nuclear expansion is required than for the scenario in which the industry grows only slowly or remains at roughly today's capacity. In any case, certain issues appear to require resolution, or significant progress towards resolution, if nuclear power is to be available as a practical response to an upswing in demand. The possibility of all three scenarios occurring simultaneously in different parts of the world cannot be discounted.

It follows then that the 'nuclear industry' needs to review its own operations in order to make them more attractive to the conditions likely to pertain in the twenty-first century. Whether 'more attractive' means 'sufficiently attractive' is not a question that can be meaningfully approached today; this project will not attempt to do so.

## 3.2 Keeping the nuclear option open

In the inevitable absence of certainty about the attractiveness of nuclear technology in comparison to alternatives in, say, 2050, the discussion should move towards 'keeping the nuclear option open'. This question could perhaps be approached by considering what would be the boundary conditions for a resurgence in nuclear power, say in thirty years.

Although the phrase 'keeping the option open' is often used, it is not clear precisely what it means (see the discussion in section 1.3). It could range from no activity, through R&D only and modest industrial activity, to expansion. Furthermore, the extent to which the option is kept open in one country is dependent on activities in other countries. For example, should there be a phase-out of nuclear power in western Europe over the next twenty years, it would be considerably easier to restart nuclear construction, say in 2050, if a complete and modern reactor design could be bought from China or India than it would if those countries had also phased out their nuclear industries.

A key issue is the length of timescales necessary in the planning of energy systems. A decision to build a nuclear station today with existing technology would not produce an operating source of electricity for at least five years, and quite possibly ten or more. Development of new technologies may take twenty years, even if work is started today.

It is likely that a certain basic supply of graduates in nuclear technology and related disciplines such as radiology, and perhaps other infrastructure issues such as nuclear laboratories, would have to be available if the option is to remain open in a meaningful sense. Such requirements would need to be quantified. At present in many countries relatively few undergraduates or graduates are opting for nuclear engineering courses, and many such courses are closing in a number of developed countries. However, many companies operating nuclear facilities seem increasingly to prefer to take more generalist engineers and train them in-house for specific nuclear activities. This necessitates retaining wide skills bases within the companies themselves. Relevant skills include not only those directly associated with the technology itself, but also an understanding of how human systems develop and interact with that technology. Such understanding could probably only be maintained within an operating system, raising questions of how to retain these skills within what is in many countries a contracting industry.

In addition, progress on a number of the issues considered in this paper would seem to be essential for the option to be open in a practical sense. Public confidence must be maintained, and in some cases improved significantly. A clear strategy for dealing with radioactive waste would also seem a prerequisite, partially but by no means entirely for reasons of public perception.

## 3.2.1 The nuclear option: the pro and anti positions

*PRO*

In order to keep the nuclear option it will be necessary to retain at least a skills base, a research community, nuclear laboratories, etc. The best way to retain the base might be to build demonstration units for new reactor and other concepts. The industry must be prepared to respond to innovation from outside the industry as well as within.

*ANTI*

For any particular country, the nuclear option will always remain 'open', in the sense that new technology can be bought from abroad or, in the absence of a world industry, developed afresh. The costs of a more active policy, particularly involving new plant, would not be justified in view of the unlikelihood that nuclear power will be needed in the future. Resources should be concentrated on more promising possibilities such as renewables and energy efficiency.

## 3.2.2 Questions for further consideration

● What would the concept 'keeping the nuclear option open' mean in practice, with reference to each of the three scenarios introduced in section 1.3?

● In particular, what would be the boundary conditions necessary for global nuclear expansion over the next thirty years?

● How serious are concerns about retention of skills for operating nuclear facilities within developed countries?

● Are there any other examples of technologies being retained in the absence of short-term demand?

## 3.3 The relative economics of nuclear power

It would seem obvious that, although matters such as public perception, waste management, etc. will be of key importance in determining the future of nuclear power, nobody will wish to build nuclear stations unless there is a design which is attractive to decision-makers, particularly in the capital markets.

Economic competitiveness is a complex issue. When considering new capacity in particular, it is difficult to make precise comparisons of 'the costs' of one energy technology against 'the costs' of another, for a variety of reasons. Clearly the cash flow patterns of necessary investment vary from fuel to fuel: some, like nuclear power and renewables, are highly capital-intensive but relatively cheap to run, while others such as combined cycle gas turbines are much cheaper to install but more expensive to operate. Simply changing the required rate of return on capital, or changing the effective pay-back period demanded by investors, can have significant effects on the relative economics of sources of electricity within these two categories, without making any changes to the costs of the technologies themselves.

In addition, factors such as geography, scale and fuel availability can have a major influence on costs. For example, some applications of renewable technologies are clearly economic now – niches such as solar-powered calculators, or wave power schemes in isolated island communities where provision of grid electricity would be prohibitively expensive. Larger-scale application such as windfarms can appear economic in certain circumstances. Distance from coalfields and quality of transportation infrastructure can have a profound influence on the economic attractiveness of coal-fired stations, a major issue in countries such as India in particular, where half of all rail capacity is at present filled by coal transport. Significantly expanding coal-fired electricity in India would therefore involve investment not only in the power stations but also in major new railway capacity. The existence or otherwise of electricity grids can make an obvious difference to the most cost-effective way of producing energy and power for relatively less industrialized communities. The potential between the best and worst wind or tidal sites in a country can be enormous in terms of quality of energy source and hence costs of production.

### 3.3.1 Changing priorities in energy investment in developed countries

Energy demand in the developed world is not expected to grow at high rates in the future. Although the share of electricity within the general energy mix is expected to increase, many developed countries have something of an over-capacity of electricity plant at present. Demand for new plant may therefore be modest in the foreseeable future and may be limited largely to replacement capacity.

In the 1970s decision-making about new nuclear stations and other generating plants lay fundamentally with governments and government-appointed regulators. Even in some countries such as Japan in which nuclear stations were in the private sector, governments regarded it as their role to ensure that strategic forms of electricity production were constructed as a hedge against any future increase in hydrocarbon fuel prices or interruptions in availability. Governments offered utilities a highly regulated local geographical monopoly, thereby creating a long-term guaranteed market for their output. As a result, they could invest in highly capital-intensive forms of electricity production (with the approval of regulators), knowing that the investment could be recouped from captive customers over a long period of time. Especially in the United States, nuclear power was regarded as being the cheapest option available.

A number of factors may have been influential in leading both to a reduction in the costs of fossil-fuel-generated electricity and to liberalization of electricity supply systems in many developed countries. They include:

- the easing of the world hydrocarbon fuel situation through the 1980s (and in particular the discovery of enormous reserves of natural gas);
- the development of the CCGT with its high thermal efficiency and lower costs;
- changing political fashion.

In liberalized markets, competing companies generating electricity have become more concerned with short-term returns on capital. Reduction of economic risk has assumed a high priority. Gas-fired capacity has become the preferred option, mainly because of its low capital costs and modular possibilities, but also because of lower perceived overall ('levelized') costs; longer-term sources such as coal and nuclear power have fallen out of favour. Gas, where available, also generally faces less public resistance. No nuclear order placed in the United States since 1977 has been completed, and building of nuclear stations and ordering of new nuclear capacity in the European Union has halted, except possibly in France.

A further, related, effect of liberalization has been to make smaller generating units, of a few hundred MW or smaller, more attractive than the 1,000 MW plus units typical of the 1970s, although some large gas-fired units, requiring relatively low capital outlay, are still being ordered. More localized generation has a number of apparent advantages over centralized generation, including:

- reduced needs for transmission over long distances;
- closer matching of load to generation;
- lower capital costs for smaller, modular units;
- less likelihood of overwhelming local pollution;
- potential for lower emissions than centralized facilities, e.g. combined heat and power;
- compatibility with decentralized political structures.

Large units face problems both in raising capital and in matching changing loads in markets increasingly characterized by a large number of small generating units. At present there is no up-to-date proven design of electricity-generating nuclear power plant of capacity below about 500 MW, although many countries, including Russia and China, are developing such concepts (Gagarinski, 2000b). Both renewables and small-scale gas turbines, including combined heat and power stations, may be better suited to a market in which an increasing proportion of the total load comprises large numbers of small users rather than a relatively small number of large customers such as, in the UK, the former Area Boards. As the average unit size falls, so it may become more difficult to manage large units coming on and off load, for example for maintenance. The likely growth of small, largely self-contained local grid connections may exacerbate this problem.

This said, baseload generation is likely to remain very important in less liberalized markets, and will still be required to some degree in more liberalized markets. Large nuclear units therefore may well remain attractive in countries such as France and Japan as well as in several developing and newly developed countries where electricity demand is growing at a fast rate. Further, liberalization has exerted downward pressure on the costs of operating existing nuclear plants, pressure that tends to be absent or much smaller in centrally controlled economies without competition. As a result many US utilities, for example, are now actively considering plant life extension, and the market for taking over nuclear reactors is becoming much more competitive. Commercial pressures also promote the introduction of new, more efficient technologies for plant operation. These would presumably be incorporated into new plant design, possibly making it more attractive.

A return to a more centralized model of electricity markets, perhaps because of resurgent fears about global energy security, or because market instruments fail to deliver sufficient reductions in greenhouse gas emissions, or because liberalized markets prove incapable of ensuring security of electricity supply on an hour-by-hour basis, might lead to a renaissance in large plants in other developed countries. Alternatively, liberalization may proceed further in countries where it is still relatively less advanced. Either of these trends might lead to a narrowing of the gaps between policies in various developed countries. It is, however, perfectly possible that major differences in nuclear policy between developed countries might persist.

### 3.3.2 The influence of the environment

Growing concerns about the environment, and especially about climate change, are leading to a reassessment of energy policy. It seems likely that a variety of market instruments to internalize the economic costs of emissions of pollutants such as carbon dioxide will be introduced. These may include:

- pollution taxation (or its less efficient surrogate, energy taxation);
- tradable emission permits;
- schemes whereby investment in pollution-saving devices in other countries can generate credits for companies carrying out such investment (Joint Implementation and the Clean Development Mechanism in the Kyoto Protocol).

Some measures have already been introduced in some countries, e.g. Norway's carbon tax (Norwegian Ministry of the Environment, 1999). Such mechanisms would improve the economics of nuclear power and renewable methods of electricity generation relative to that of fossil fuels. Pollution taxation would, however, do little to overcome the inherent economic risk associated with large-scale highly capital-intensive projects; smaller reactor models, some of which are in an advanced state of design (see section 3.4), may still be required in some markets.

### 3.3.3 The developing and newly industrialized world

At present the richest 20 per cent of the world's population use 55 per cent of the world's energy, while the poorest 20 per cent use only 5 per cent (WEC/IIASA, 1998). Although it is unlikely that such disparities will disappear rapidly, it is at least likely that most of the expected growth in world energy (and especially electricity) demand over the next fifty years will be in countries which are at present relatively less developed.

The situation in many developing countries is rather different from that in the market economies. In many Asia-Pacific countries, for example, demand for energy continues to grow rapidly, but access to plentiful oil and gas reserves or convenient coal reserves is, for the present at least, limited. As a result, several countries, such as China, India and South Korea (as well as Japan), remain publicly committed to large-scale development of nuclear power in the immediate future. Significant energy growth is also expected in Africa and South America, where (with the exceptions of South Africa, Brazil and Argentina) nuclear power is not at present used.

These economies retain a degree of central control, which makes raising capital for large-scale schemes less problematic than in the more market-oriented systems. Further, the rate at which energy (and especially electricity) demand has been growing, for example in China, makes it unlikely that any single source of electricity can expand sufficiently rapidly to satisfy it. China has enormous coal reserves, but these are largely to be found in the north and northwest of the country, while demand is growing most rapidly in the southeast, and the rail infrastructure is incapable of carrying major increases in supply in the short term. China is also concerned about the local and global environmental impacts of continuing rapid expansion in its use of coal.

However, questions remain over the suitability of nuclear power for some developing nations. Capital is often a constraining factor, especially in view of the current unwillingness of many world financial institutions to lend money for nuclear projects. Large-capacity plants, such as modern nuclear stations, require a major grid infrastructure to be effective. It is argued that the complexities of nuclear science and technology compared to lower-technology fuels would offer challenges to the engineering infrastructure of some developing (and indeed developed) countries, although they have proved quite able to handle many complex technologies such as aviation. There is an associated need for ongoing training and R&D support. It has sometimes been suggested that developed nations should increase their use of nuclear power to allow more use of fossil fuels in developing countries, but it is difficult to imagine a global political and economic mechanism whereby this could be achieved even if it were deemed appropriate.

Voices within the former Soviet Union also talk of a renaissance in nuclear construction, involving both the completion of abandoned projects such as Rivne-4 and

Khmelnitskiy-2 in Ukraine, and new plants. However, the situation within the FSU is volatile. The continued operation of RBMK reactors, the design used in the Chernobyl plant, at sites in Lithuania and Russia represents a potential threat to nuclear revival. The VVER, the Russian version of the pressurized water reactor, is recognized as a much better design. However, there is considerable concern not only about the safety of early VVERs in eastern and central Europe, but also perhaps about the levels of regulation surrounding exported VVER plants to a number of developing countries. Russia appears to remain committed to reprocessing, and possibly to fast reactors.

It is possible, then, to envisage a future in which some countries, perhaps those which retain a regulated electricity supply market, will continue to develop their nuclear industries, while nuclear power construction in some of the market economies may be limited to replacement capacity, or perhaps not even to that.

### 3.3.4 The economic argument: the anti and pro positions

| ANTI | PRO |
| --- | --- |
| Market liberalization has revealed the fundamentally uneconomic nature of nuclear investment. Unlike most forms of energy, nuclear power has become more expensive as time has passed, and this trend is likely to continue. There is | Nuclear power can be economic, and indeed in some areas already is, at least at rates of return below about 8 per cent (IEA, 1998). At present it is the only major source of electricity that accounts for its waste management costs. When |

no reason to believe that fossil fuels will sustain high price rises in the long term: history suggests such prices remain remarkably constant in real terms, though with short-term variations.

fossil fuels have to do the same, either through pollution taxation or through tradable emission permits, relative nuclear economics will look more promising, especially if fossil fuel prices rise as more easily extracted reserves become exhausted.

The problems of nuclear power are such that there should be no government subsidy, hidden or overt, to aid its development.

Some level of government involvement will be necessary in areas such as regulatory stability and insurance against major accident.

Nuclear power is particularly unsuitable for developing nations, in terms of both the difficulty of raising the necessary capital for a large programme, and the safety challenges it offers. All attempts should be made to make alternative energy sources available to such countries.

Some developing nations are likely to invest in nuclear power whatever the stance of the developed world. A vigorous nuclear industry in the developed world would allow for technology transfer with less developed countries, hence assisting their development and improving safety standards.

## 3.3.5 Questions for further consideration

- What are the relative economics of nuclear energy against alternatives, taking the full cycle into account and considering the effects, for example, of different rates of return, future fuel costs, costs of waste management, etc.?

- What factors will affect the relative attractiveness of large and smaller reactors in developed and developing countries?

- How is liberalization of electricity supply markets affecting investment decisions?

- What ongoing role will be required of governments in liberalized markets?

- What would be the effects of introduction of market instruments such as tradable emission permits or pollution taxation to protect the atmosphere?

- Why have nuclear costs tended to increase during a period when other methods of generating electricity have become cheaper?

## 3.4 Nuclear research and development

Research and development is a key issue in addressing the question of the boundary conditions in which nuclear power may have a role to play over the next decades. As suggested in 3.1 above, one can imagine circumstances propitious to a return to nuclear technology, should the need to stabilize the greenhouse gas content of the atmosphere by mid-century be confirmed, or if questions over energy supply for the rapidly growing markets of the developing world were to pressure governments to increase controls over electricity markets. Indeed, current circumstances in many developing countries exhibit these features today. However, even if circumstances should prove ripe for a growth in demand for nuclear power, it is not certain that today's technology would be appropriate in all countries. Large, highly capital-intensive units might prove unsuited to mature electricity markets, and smaller units might prove necessary (although, alternatively, large units might prove necessary for a stable and competitive generating system in developed countries). Increased public concerns about nuclear power would also have to be overcome, as these have been exacerbated by incidents such as the accidents at Three Mile Island in 1979 and Chernobyl in 1986. If radical new approaches to reactor design are required, perhaps to run alongside existing approaches which may remain attractive in less liberalized markets, then considerable research, development and demonstration efforts will be needed.

The availability of uranium is also an issue. Despite early projections about the supposed scarcity of uranium and therefore the need to develop a system of breeder reactors, most nuclear power in the world is at present produced in reactors running on uranium, either in a natural state or modestly enriched in the fissile isotope uranium-235, which constitutes 0.7 per cent of natural uranium. Some reprocessing is taking place, and the resulting separated plutonium is being offered for use in thermal nuclear stations as mixed oxide (MOx) fuel; there is also a small number of research and prototype fast reactors (plus the Russian power reactor BN600) fuelled at least in part by plutonium. However, use of plutonium currently represents only a very small proportion of world nuclear generation.

It is generally stated that, at current rates of usage, uranium for use in thermal stations will last about a century, although it would seem reasonable to assume that this reserve, like those of other fuels, would prove to be much larger should demand return. Other estimates (e.g. Holdren and Pachauri, 1992) suggest that land-based uranium reserves might last for over 300 years at current rates of usage, while inclusion of uranium in seawater would stretch the reserve-to-usage ratio to over 50,000 years. In addition, thorium, a significant reserve, may make a contribution, although there are different views as to its technical suitability and resistance to weapons proliferation.

A nuclear industry of about the current size, based on replacement capacity and perhaps modest growth, could probably continue in its present mode for quite some time, although other issues would have to be addressed. Reprocessing might not therefore be necessary to increase fuel reserves, though it could be attractive for other reasons, e.g. reducing the volumes of fresh uranium to be mined (and therefore local environmental damage) and waste management issues.

However, should nuclear generation expand significantly, say tenfold, the strain on world uranium reserves would increase, even if the expansion were based on reactor designs with rather higher uranium burn-up factors than at present. It is not certain that a once-through system based on natural uranium (and perhaps thorium) could meet the increased demand.

R&D might demonstrate that extracting uranium from seawater is a commercial proposition, with lower fuel cycle costs than those associated with a reprocessing route. If so, a thermal reactor-based programme of nuclear stations, using a once-through approach to fuel, could feasibly form the basis of a much larger world nuclear industry for some centuries. Alternatively, reprocessing might become an essential element in a significantly larger nuclear industry, should reserves of once-through fuel (uranium and possibly thorium) be limited. The plutonium so separated could be used in thermal reactors as MOx, or in fast reactors. The implications of this are explored in section 4.5.

The conclusion would seem to be that, if the nuclear power option is to be available to respond to several possible futures (including the three scenarios introduced in section 1.3), major research and development may have to be directed towards areas such as spent fuel management (e.g. partition and transmutation), the thorium cycle, reactors with lower capital costs, smaller outputs and passive safety features, new methods of extracting uranium say from seawater, new reprocessing techniques, etc. Clearly, not all, if any, of these developments would necessarily be required in any particular scenario, but R&D into all might be regarded as wise, given the uncertain nature of energy futures. The question then arises of who should fund such a research effort.

### 3.4.1 R&D in liberalized electricity supply markets

The liberalization of energy supply industries in the developed world has had a marked effect on R&D expenditure, especially on projects with medium-term or long-term projected pay-back.

In the 1970s, when supplies of hydrocarbon fuels appeared under considerable threat and reserves were thought to be limited to a few decades, government spending on R&D was high in most developed countries, either through direct programmes or through state-owned energy utilities. Government believed its role included the development (and indeed deployment) of alternative sources of energy to reduce dependence on imported hydrocarbons, especially from the Middle East.

As the oil price fell and new reserves of gas were discovered, coupled with development of the combined cycle gas turbine (from 1980 onwards), so the urgency to develop alternatives declined. Further, as governments privatized and liberalized energy supply industries, so attitudes to state-funded R&D changed.

In effect, a different answer was being given to the question, 'Who benefits from energy R&D?' Heavy regulation of electricity supply systems was instituted in many countries partly because of fears that a free market could not deliver secure and stable electricity supply systems in the long term for a variety of possible reasons. The same motivation led governments to carry out large programmes of energy R&D. It was perceived that the interests of the electricity consumer (in effect, the taxpayer) could not be safeguarded without considerable state involvement, both in ongoing operational matters (through regulation of the delivery systems) and in the longer term (through R&D). Since the taxpayer would ultimately benefit from these measures, it was deemed appropriate that the taxpayer should pay for them.

As fears about long-term energy supplies diminished, however, it was increasingly held (or at least articulated) by governments that R&D which would ultimately offer

**Table 2: US Department of Energy spending on applied energy technology R&D**

| Year | Spend (1997 $US billion) |
|------|--------------------------|
| 1978 | 6.15 |
| 1992 | 2.18 |
| 1997 | 1.28 |

financial benefits to privately owned companies should be undertaken by those companies, not by the taxpayer. The direct beneficiaries of energy R&D would be the energy companies, so they should pay for it. In many developed countries state expenditure on energy R&D fell dramatically, following the easing of the world oil supply situation in the mid-1980s. For example, US Department of Energy spending on applied energy technology R&D has changed as shown in Table 2.

As a fraction of GDP, the USDOE's spending on energy technology R&D in 1997 was less than half that of its predecessor agencies in 1967 when oil prices were at a historic low. In 1997 nuclear fission R&D amounted to 3.7 per cent of that of 1978, and even renewables R&D was running at only 18.5 per cent of 1978 levels (OCFO, 1997). At the same time, for reasons explored below, long-term R&D spending by companies fell in those countries which were in the process of liberalizing their electricity supply markets. In the United States, for example, industrial expenditure on energy R&D fell from an estimated $4.4 billion in 1985 to $2.6 billion in 1994 (Dooley, 1994).

## Table 3: Government energy R&D budgets (US$ million, 1995 values)

|      | Canada | France | Germany | Italy | Japan | Sweden | UK  |
|------|--------|--------|---------|-------|-------|--------|-----|
| 1985 | 469    | 714*   | 1589    | 1137  | 4335  | 162    | 708 |
| 1995 | 239    | 672    | 358     | 290   | 4714  | 64     | 83  |

* 1990 figure.
*Source:* IEA, 1997.

Similar patterns have been seen in most developed countries. Between 1984 and 1994 public-sector energy R&D fell (in real terms) more than fourfold in Germany and Italy, eightfold in the UK, twofold in Canada. It is noticeable that state-sponsored R&D has continued at higher levels in countries where liberalization of electricity supply markets is relatively less advanced. The rate of reduction in France was somewhat slower than in other west European countries, but the only real exception was Japan, where it grew from US$1.4 billion in 1974 to $3.9 billion in 1980 and $4.7 billion in 1995 (in constant dollar terms) (see Table 3).

In a competitive energy market the relationship between companies, governments and shareholders is different from that in a heavily regulated system. Shareholders, to whom companies are now primarily responsible, tend to be more averse to economic risk than governments. Partly because of the level of discount rate used by commercial energy companies, companies generating electricity have proved unwilling to indulge in some of the long-term speculative research that was previously supported by governments.

Research continues to be attractive to the vendors of nuclear technology, but the extent to which such companies are able to fund 'revolutionary' nuclear research (as opposed to evolutions of today's technology) within a contracting world market is bound to be limited. This has been especially so for research requiring large demonstration facilities, although R&D into matters such as operational performance, plant lifetime extension and some longer-term 'small science' projects, such as fuel cells, has continued. Contrary to the declared assumptions of some governments, state-funded and privately funded R&D expenditure tend to rise and fall together, rather than a reduction in the one being compensated by an increase in the other. R&D has become increasingly packaged in smaller projects, aimed at incremental improvements in existing technologies, rather than at 'breakthrough' demonstration projects.

The issue is exacerbated by the relative unattractiveness of capital-intensive sources of energy such as nuclear power in competitive markets, all else being equal. Companies will be even less attracted to sponsoring R&D in fields where the technology itself looks unattractive, with the exception of issues associated with lifetime extension and operation of existing plants.

The market-driven move away from big technologies, both in R&D and deployment terms, is exacerbated by perceptions of failings in previous nuclear R&D projects, notably during the 1960s and 1970s.

## 3.4.2 An inescapable role for government

It appears to be relatively uncontroversial that governments have an important role in setting the basic parameters within which industry should operate, and in protecting and promoting suitable infrastructure, on behalf of society as a whole.

It looks unlikely, for reasons adduced above (e.g. in sections 1.3 and 3.4), that the present nuclear reactors and infrastructure will be appropriate for all possible future scenarios. New approaches may well need to be developed. However, it also seems unlikely that research funded solely by individual private companies could develop smaller reactors, new ways of approaching radioactive waste, etc. A demonstration plant for a new reactor design, deep repository or transmutation assembly would be expensive. Similarly, although the thorium cycle is well characterized in the literature, it would require a new infrastructure to support it. Governments, like private companies, may prefer to focus R&D efforts on technologies which are likely to progress in relatively small incremental steps (where such options are available) rather than requiring large pilot or demonstration facilities. It is not clear that new nuclear technology could proceed in such a way, while improvements in many renewables, such as wind power or solar power, may do so.

However, attitudes may be changing. Bearing in mind that in the energy industries the gap between initial research and widespread commercial deployment of new energy sources can be a matter of decades, pressure on governments seems to be growing to increase research and development expenditure in this area. The US federal government, for example, spent $1.2 billion per year on renewables in 1999/2000. It is accepted that the private sector alone is unlikely to carry out long-term fundamental research, although development of sources close to commercial exploitation may well be attractive to major energy companies (e.g. BP and Shell Solar). Many respected bodies, such as the US Presidential Committee of Advisors on Science and Technology (PCAST, 1999) and the Royal Society in the UK (Royal Society/ Royal Academy of Engineers, 1999), have recognized the role of governments and argued for an inclusive approach to research into non-greenhouse-gas-emitting energy technologies, with a view to developing a portfolio approach to combating climate change.

In effect, governments may be accepting once again that there are elements of R&D, beneficial to taxpayers in general, which are unlikely to be carried out without government involvement. However, in many countries, authorities funding research are offering grants only when industry is contributing a similar level of resources. An example is the EPRI (Electric Power Research Institute) in the United States, an association of the research arms of electricity utilities which actively seeks involvement from government and research institutes and has an international presence. This approach might be thought likely to combine assurance that new techniques will be developed with market deployment in mind, while recognizing that the greater public good is safeguarded by taxpayer-funded R&D.

In major technologies such as nuclear power, R&D progresses in three broad phases – the science, the engineering (leading to the building of pilot plants) and the economic (leading to the building of full-scale demonstration plants). The scientific phase is usually the least expensive, but often takes the longest time. Some commentators have suggested that investment in the science of new concepts (e.g. passively safe reactor designs with lower capital costs; partition and transmutation; extraction of uranium from seawater) could be carried out over the next few years. Decisions on building pilot plants and, ultimately, demonstration facilities could then be left until more information is available about likely attractiveness.

In fact, the International Atomic Energy Agency (IAEA) has an inventory of some thirty innovative and evolutionary reactor designs which are at various stages of consideration in various parts of the world. At least two designs are well advanced, and appear close to demonstration status. The Westinghouse AP-600 (a much simplified 600 MW water-cooled reactor) received final design approval from the US Nuclear Regulatory Commission in 1998 (Westinghouse, 2000). The ESKOM (South Africa) Pebble Bed Modular Reactor (PBMR, a high-temperature gas-cooled reactor-based system with units of 110 MW output – ESKOM, 2000) has attracted international interest and support. Both these designs claim shorter construction times and lower capital outlay per unit of electrical output than present reactor designs. Such concepts are not entirely new, and it would be premature to assume success in any individual case. It may be noted that many novel reactor concepts are more in the 'development' than the 'research' phase, and it is the development phase that is generally the more expensive. Certain new initiatives, such as the Generation-4 reactor programme, are being followed, with the aim of minimizing development costs by leveraging strengths available throughout the world.

A further move, which might increase the attractiveness of long-term R&D for governments and the private sector alike, would be a greater degree of international collaboration. Some international projects, such as those carried out by the European Union on renewables and energy efficiency, have already been successful. Examples in the nuclear field include the joint European projects on nuclear fusion (JET, sited in the UK), high-temperature gas reactors (Dragon, again in the UK) and waste repositories at Äspö in Sweden and Grimsel in Switzerland. International collaboration between private companies, especially in association with governments, could similarly achieve progress outside the reach of any individual company.

However, the issue of 'who benefits' is complex. It may be, for example, that developing countries might benefit the most from new electricity technologies, as they have the fastest-growing markets. It would be unlikely, however, that developing countries could contribute a major, or even a pro rata, share of the R&D funding. Nonetheless, in environmental terms it can persuasively be argued that reductions in carbon dioxide emissions, wherever they occur in the world, will benefit taxpayers in developed countries. The governments of developed countries should therefore be prepared to contribute towards international collaborative research and development initiatives, be they renewable, nuclear or carbon sequestration in nature.

This said, R&D funded by governments has often been relatively poorly controlled, sometimes appearing to be research for research's sake rather than being aimed at a specific result. It has often lacked a focus on deployment and therefore represented poor value for money. As suggested above, match-funding of projects between government and the private sector is one possible approach to this problem.

### 3.4.3 Nuclear R&D: the pro and anti positions

| PRO | ANTI |
| --- | --- |
| The competitive structure of electricity supply markets makes it difficult for organizations within them to invest in long-term research that, like all research, risks being unsuccessful. Lessons have been learned from the approach of the 1960s and 1970s, in which nuclear R&D was often technology-driven and represented poor value for money. Nuclear R&D in the twenty-first century will be more market-led, and hence more likely to produce workable solutions. The problems to be overcome – especially the issue of capital costs – are more closely defined than they were in the 1960s. | Nuclear power was the recipient of enormous research funds from the 1950s to the 1990s. This research produced a source of energy responsible for just 7.5 per cent of world-traded primary energy. There is no reason to believe that a new research effort now would be more successful. State-funded research and development efforts should be directed towards renewable technologies and energy efficiency; the nuclear industry itself should be responsible for ongoing nuclear research. The priorities should be waste management and decommissioning rather than new reactor systems. |

A major expansion might
require the development of
new approaches to nuclear
power. Such a research effort
cannot be met solely by
private companies.

---

### 3.4.4 Questions for further consideration

- Under what circumstances might R&D into radical new reactor concepts and/or fuel cycles be attractive to governments?

- What types of nuclear technology would be required to support the different scenarios in section 1.3?

- What mechanisms are available for funding long-term energy R&D (e.g. joint ventures, industry-led schemes, international collaboration) and what are their relative merits?

- What would be the cost of R&D to develop smaller, more flexible nuclear designs?

- On what basis should R&D resources be allocated among various potential methods of reducing greenhouse gas emissions?

## 3.5 Skills required if nuclear power is phased out

One of the three scenarios introduced in section 1.3 covers the slow phase-out of the nuclear power industry over the next five or six decades. Many companies make provision for back-end costs, generally by reinvestment in the ongoing business. Investment in new plant includes an implicit provision for whatever environmental and other costs may arise in the future, such costs appearing on the balance sheet. There are few examples of companies making such provisions for periods of 100 years or more.

However, an industry which may no longer be investing in cash-generating assets is in a very different position. Either provision has to be made from its current operations, or at some point the state will need to make suitable provisions (e.g. dealing with contaminated water from disused coalmines).

At present, there is a variety of methods in place in various countries for ensuring at least partial provision of funds for eventual waste management, decommissioning, site decontamination, etc. All take as the fundamental principle that the polluter should pay.

In some cases, nuclear operating companies pay money into a segregated fund for discharge of future liabilities. (The US waste management fund, comprising a levy on nuclear electricity and held by the federal government, is an example. In Sweden and Finland responsibility for taking care of decommissioning and waste management remains with the owner of the operating licence, as well as responsibility to provide, during the operating life of the facility, sufficient funds for these activities.)

Some of the issues affecting the skills required to ensure safe long-term management of nuclear liabilities are also relevant to the topics surrounding 'keeping the nuclear option open', and those concerned with waste management. In reality, even if no further nuclear stations were to be constructed, there would be a necessary programme of work stretching over many decades. The skills involved would be focused on waste management and on decommissioning and decontamination of sites. R&D work into better technologies for dealing with these issues would remain a priority.

Perhaps more likely than simultaneous phase-out of nuclear power in all countries would be a situation in which some countries were actively or passively moving away from nuclear power, while others were continuing to invest and perhaps even expanding their capacity. In such a scenario it is likely that leadership in world nuclear matters would shift from its current locus in the developed world towards countries, perhaps in the Asia-Pacific region or in central and eastern Europe, which continued to invest in nuclear technology.

Phasing out nuclear power could have significant implications for non-power uses of radioisotopes, including medicine. Unless such techniques were to be abandoned at the same time (perhaps because alternatives not depending on ionizing radiation were developed), there would be an ongoing requirement for materials, irradiation facilities, waste management programmes, etc., but a much weaker civil nuclear infrastructure to support them.

Similarly, it would seem likely that the implications of a phase-out of civil nuclear technology would differ depending on whether there was a simultaneous phase-out of nuclear weapons. A number of countries have shown interest in nuclear weapons without a significant nuclear power programme – Israel, Pakistan, China (which has only recently started to deploy nuclear power), North Korea. Some of these countries, and the other nuclear weapons states, might continue to deploy nuclear weapons, and hence face issues about nuclear reactors, waste facilities, etc., even should nuclear power be abandoned.

### 3.5.1 Nuclear phase-out: the pro and anti positions

PRO

The attractions of nuclear technology are such that early phase-out would be foolish. Some level of ongoing nuclear technology, for medicine, research, industry, agriculture, etc. and quite possibly for military uses, is likely to continue, even should civil nuclear power not thrive.

There will also be a continuing need to develop better technology for decontamination and management of waste.

ANTI

Nuclear power should be phased out as soon as possible. However there will still be many decades of work needed to clean up the nuclear legacy. The nuclear industry should be required to establish segregated funds to provide for this work when necessary. Opportunities for the transfer of useful technologies to other industries should be evaluated.

### 3.5.2 Questions for further consideration

● What skills will have to be safeguarded and improved should nuclear power be phased out globally, and how could this be achieved?

- How would the issue be affected were nuclear weapons to be retained, or not retained, after nuclear power was phased out?

- If nuclear weapons are not retained, what would be the effect of phasing out nuclear power on the skills required to maintain military nuclear facilities and those associated with other uses of radioactive materials, such as medical uses, including their eventual clean-up?

- What concerns, if any, would be raised by a shift of world leadership on nuclear issues from developed counties to developing countries, should nuclear power continue to decline in North America and western Europe?

- What models of provision for future liabilities in the absence of a cash-generative industry might be most effective?

- What could and should be done to the world's stock-piles of highly enriched uranium and plutonium, including redundant military materials, in the absence of a civil nuclear programme?

# 4   ISSUES OVER WHICH THE NUCLEAR INDUSTRY HAS PARTIAL INFLUENCE

## 4.1  Introduction

It is in the nature of some issues that, although the actions of the nuclear industry will have some importance on them, so too will circumstances outside the industry's control.

## 4.2  Nuclear proliferation

Nuclear weapons predate civil nuclear power. It took less than three years, using 1940s technology, from the first demonstration of nuclear fission to delivery of the first atomic weapons. It is therefore highly improbable that any purely technical steps could be found to prevent a determined state from developing at least fission weapons. It would appear unlikely that a state such as South Africa, which has renounced nuclear weapons and destroyed its capabilities, would restart a military programme simply because it developed the Pebble Bed Modular Reactor, for example. However, it is also likely that states with access to research reactors and a base of nuclear engineers through involvement in nuclear power might find it somewhat easier to develop a military capability should they so wish.

Fears of proliferation are not uniform, the matter seeming to be of more concern in the United States than in Europe (as reflected, for example, in US antipathy towards reprocessing). Such fears fall into two broad categories: fear of acquisition of weapons technologies and materials by states, and fear of their acquisition by terrorist groups.

Any major expansion of nuclear generation which involved a move towards fast breeder reactors, with consequent requirements for reprocessing, would herald the 'plutonium economy', in which much more frequent transportation of plutonium fuel might be necessary, with potential consequences for international security and, possibly, personal liberties.

Even in the absence of a fast reactor programme, an increase in thermal nuclear generation based on uranium could have proliferation implications. Increased transportation of nuclear materials, coupled with greater dissemination of nuclear technology, could increase the risk of diversion of both technology and materials to military purposes, either by states or by terrorist organizations, although this risk could be avoided if new technologies or practices were developed. The non-proliferation regime instituted by the International Atomic Energy Agency and developed through the Nuclear Non-Proliferation Treaty (NNPT) has been quite successful. However, the end of the Cold War may have increased risks of proliferation (or at least increased international concern), as states which once were clients of the United States or USSR reassess their security priorities. A major expansion of nuclear power might put considerable strain on present approaches to safeguards, and might require new regimes.

Moreover, research or materials reactors might technically be used for proliferation purposes. Indeed, the requirement for isotopic separation when producing radioactive materials for medical purposes, for example, might make this a more attractive route to potential proliferators than the civil nuclear power route. In view of the vital role such facilities play in nuclear medicine, non-destructive testing, etc. it seems unlikely that they would be phased out even if use of nuclear power were to decline. They can also be built quickly – the first Hanford production reactor was constructed in four months. Some institutional approach to non-proliferation would presumably be required indefinitely.

It is possible to envisage a scenario in which military nuclear technology is abandoned or significantly scaled back, while civil nuclear power continues to be used in at least some countries. Managing the uranium and plutonium liberated by the destruction of nuclear weapons would then be a major issue, and indeed is becoming so already. One technical route could involve burning up military plutonium and uranium stockpiles in thermal or fast reactors.

One further aspect of the use of nuclear materials for military purposes which has an impact on perceptions of civil nuclear power is the treatment and disposal of waste from military uses. These include contaminated weapons sites and, of special concern in the former Soviet Union, management of redundant nuclear submarines, most of which are at present stored with their reactors *in situ*.

It can be argued that not separating plutonium from spent nuclear fuel would significantly reduce, if not eliminate, the risk of terrorist groups developing weapons technology, as extraction of plutonium from fresh highly radioactive spent fuel is difficult and dangerous. However, after some decades of cooling, such dangers are much reduced, so that states could develop or use small-scale methods of extracting plutonium from spent fuel. Spent fuel in dry storage or in a deep repository would therefore represent some proliferation risk.

## 4.2.1 Nuclear weapons proliferation: the anti and pro positions

| ANTI | PRO |
|---|---|
| Increasing international transportation of nuclear materials, which would be necessary in an expanding nuclear industry, would lead to an unacceptable increase in the likelihood of materials being diverted or stolen for military purposes. The creation of centres of nuclear expertise, should civil nuclear technology spread, would make it easier for the countries where they are located to develop a military capability. | Nuclear weapons predate civil nuclear technology. There are no 'technical fixes' to weapons proliferation. The offer of aid in developing civil nuclear technology in return for guarantees that materials and technology will not be diverted towards military purposes has been successful in preventing proliferation. Removal of this aid and hence of the incentive to renounce nuclear weapons would weaken anti-proliferation measures. Civil nuclear power offers a possible means of destroying stockpiles of nuclear materials arising from destruction of nuclear weapons. |

## 4.2.2  Questions for further consideration

- Are the IAEA and the NNPT as currently constituted and funded still suitable foci for non-proliferation work under all three scenarios presented in section 1.3.4?

- What are the implications for proliferation of the growing use of plutonium for civil purposes?

- Are more proliferation-resistant fuel cycles feasible?

## 4.3  Public perceptions, politics and decision-making

In discussing public attitudes we find ourselves stepping into the realms of human values, theology and even symbolism. Nuclear power and radioactivity have to some extent taken on the mantle of science and technology as a whole, and may even have resonances with matters deep in the collective unconscious, such as the Frankenstein myth, alchemy and the Fall of Man (Weart, 1988). For some people technology represents an enormous benefit whereas for others it is a symbol of humankind's rejection of a more wholesome, natural way of life. It is obviously not within the scope of this study to adjudicate between different world views, even if this were in principle possible. However, it seems very unlikely that simply coming to a consensus about the statistical safety of a particular reactor design, for example, would lead to a consensus about use of an advanced technology such as nuclear power.

### 4.3.1 The role of 'facts' in guiding public perceptions

There are several examples of topics in which the 'facts' are matters of relatively little dispute but lead to diametrically opposed responses. Three Mile Island, to the nuclear industry, was a demonstration of the safety of the technology. Although 3,000 MW of thermal power went out of control, nobody was hurt because the safety systems worked. Opponents, by contrast, often cite Three Mile Island as an example of why the technology should be regarded as unacceptably dangerous. To advocates of nuclear power, plans to bury nuclear waste 1,000 metres underground demonstrate how safe the industry is; to opponents, they demonstrate how dangerous the waste is.

In areas of scientific dispute, people frequently seem to believe only those 'facts' that support their personal attitude. At the extreme, polls show that a significant number of people who class themselves as 'pro-nuclear' cannot think of a single benefit of the technology, and a similar number of those who describe themselves as 'anti-nuclear' cannot think of any disadvantage. It is likely that even the most 'objective' of observers tend to come to decisions based on their judgment of the credibility and values of the sources of information. In addition, perceptions of risk tend to be affected by perceptions of reward. In countries where the demand for nuclear power has been greatest, perhaps because there are no other domestic power resources, support for nuclear power tends to be greater. In those countries where the benefits offered by the industry are less obvious, by contrast, it has proved more difficult to maintain public confidence.

### 4.3.2 The nature of 'public opinion'

The field of public opinion is replete with apparent paradoxes. Most people seem capable of holding quite contradictory views simultaneously, for example wanting reductions in traffic congestion and pollution but opposing measures to make car travel less attractive. The NIMBY (Not In My Back Yard) syndrome is a further example. Power plants or even former weapons test sites may cause less anxiety than proposed waste sites. Sometimes local communities can support a project which is subsequently overturned at state or federal level. Furthermore, it is well established that the apparent state of 'public opinion' can be profoundly affected by the particular wording of the question asked, or even the way it is asked. Questions about nuclear power in the context of climate change and possible future fuel shortages elicit more positive responses than the same questions in the context of a major accident such as Chernobyl or 'the insoluble problem of nuclear waste'.

### 4.3.3 Perceptions of public perceptions

When the effect of public opinion on decision-making is considered, a further complication occurs. Decision-makers, naturally, base their decisions partly on their perception of public opinion, in other words on the perception of a perception. There is some evidence that these second-order perceptions may also be subject to error.

Opinion polling carried out by MORI (BNFL, 1999) in the UK suggests an interesting pattern of perceptions

**Table 4: MPs' perceptions of public opinion (%)**

|  | Favourable towards nuclear energy industry | Unfavourable towards nuclear energy industry | Neither favourable nor unfavourable/ don't know |
|---|---|---|---|
| Public opinion | 28 | 25 | 47 |
| All MPs | 43 | 44 | 13 |
| MPs' perception of national public opinion | 2 | 84 | 14 |

among opinion formers and decision-takers about public perceptions (see Table 4).

A similar pattern has been observed in the United States. These data imply that, at least in some countries, decision-takers' perceptions of public opinion may not be accurate; thus their decisions may be skewed by assumptions that may not be true. However, further paradoxes emerge. Opinion polls in Sweden, for example, show a considerable majority in favour of existing nuclear power stations continuing to operate (Analysis Group, 2000), a view shared by most of Swedish industry, yet the governing coalition forced closure of one nuclear plant in 1999, and another closure is, in principle, due to follow. The internal politics of governing coalitions may also be a factor in determining policy.

## 4.3.4 Decision-making in modern democracies

As in many other areas of the nuclear debate, it would be a mistake to expect identical, or even similar, decision-making processes to take place in all countries, even in those with apparently similar circumstances or apparently similar policies towards energy.

A survey of decision-making in the early days of the nuclear industry (like many other industries at that period) reveals it to be characterized by:

- considerable, perhaps excessive, secrecy;
- arrogance;
- decision-making involving only a small number of nuclear experts;
- the 'decide, announce, defend' (DAD) model.

It has since become clear that, although these modes of operation may have been acceptable in the 1950s and 1960s, they have led to considerable suspicion about the industry on the part of many people. In addition, decision-making in some countries, notably those with liberalized electricity markets, has become more complex simply on account of the greater number of stakeholders involved (see section 3.3). As a result, in some countries the industry has found it more difficult to move forward on individual projects, such as the siting or operation of waste management repositories and new nuclear power stations, while even some existing facilities have come under pressure. In the words of a former UK Environment Minister, DAD has been supplanted by DADA (decide, announce, defend, abandon).

It would seem clear, then, that generalized public attitudes to nuclear power are only part of the story. Building legitimacy and consensus around specific decisions and programmes is increasingly important. It is a truism to say that continued operation, let alone expansion, of the nuclear industry will require new facilities to be built as existing plants come off-line. Although current plant economics in countries such as the United States have improved so far that lifetime extension is now much advocated, there will come a time when it is no longer economic to refurbish an existing nuclear plant, and new capacity of some kind will have to be ordered.

Even proposals to build replacement capacity on the same site could encounter considerable local opposition (e.g. the Hinkley Point C proposals in the UK, which were opposed by a consortium of over twenty local councils). It has been an increasing trend in some countries, notably Canada, the UK and Germany, for protesters to take part in direct action, sometimes of a violent nature, against specific proposals or facilities in areas such as road building, luxury home development, logging, animal experimentation or nuclear transport. Such actions are often directed against individual workers or company directors in an attempt, often successful, to intimidate them so much that they abandon the plans in question.

Even in the absence of these extreme actions, local communities have proved increasingly adept at building consensus against unpopular projects, even (or perhaps especially) those with a national or federal dimension. (It might well prove even more difficult to win local consent for an international waste repository, for example, than for a national or local one.) Political actions involving local politicians have become more prevalent, and are often coupled with sophisticated local campaigns which make it difficult for opposing views to be expressed.

Statutory public involvement in decision-making has been increasing. For example, in many countries such as Canada and the United States an Environmental Impact Assessment is a regulatory requirement for the construction and licensing of a nuclear facility. Typically, such EIAs include public consultation and involvement to ensure that local concerns are addressed. Considerable work has also been done to develop innovative ways of ensuring more involvement from potentially affected communities and society at large in such decisions. Such work – it is sometimes referred to as a 'stakeholder' approach – has included citizens' panels, consensus conferences (in Denmark, the UK and the United States, for example) and stakeholder dialogues, in some cases building on the experience of well-established local liaison committees. Ways of using the planning process so that any development will benefit local communities are also being explored.

As a corollary, the decision-making processes of industry and government must become more open, and must involve groups other than technical experts at an early stage. The ultimate aim should be to foster 'consent', not merely 'acceptance', among potentially affected communities.

### 4.3.5 Public perceptions: the pro and anti positions

| PRO | ANTI |
| --- | --- |
| Public fears about nuclear power do not correlate with the actual safety record of the industry compared to other power technologies. | The public is rightly hostile towards nuclear power, given the dangers inherent in the technology. Over time people have become more anti-nuclear, as the risks inherent in the technology have become more obvious. |
| Generally speaking, the more people know about nuclear technology the less concerned they are. | Education about nuclear power tends to increase people's concerns. |
| Nuclear fears, such as they are, tend to be very back-of-mind. In most countries with nuclear stations, nuclear power does not attract a great deal of attention outside campaigning organizations. | Although nuclear fears may be back-of-mind in the absence of a particular project or programme, they come very much to the fore when actual projects are proposed. |
| The media and the more vociferous pressure groups do not represent the real state of public opinion. | The secretive and inflexible stance taken by the nuclear industry has contributed to public disquiet. |

### 4.3.6 Questions for further consideration

- What factors affect the way in which people assess the risks associated with various activities?

- How can public opinion be gauged, and how should the results of public opinion polling be interpreted?

- What is the state of public opinion in a selection of countries, and how is it changing?

- How are decision-makers' perceptions of public opinion constructed, and what influences them?

- How do the attitudes and actions of governments affect public opinion?

- How do decision-makers' perceptions of public opinion affect their decisions?

- What new models of stakeholder involvement in decision-making are likely to be effective in ensuring the early involvement of local and wider communities?

## 4.4 Waste management

A constant battle against wastes of various descriptions and types seems an inevitable feature of industrialized societies. In particular, all energy sources produce waste, whether in the course of extracting the fuel, manufacturing the machinery, normal operation or dismantling. Inevitably, then, the waste produced by any one source must be compared with the wastes of others, notably fossil fuels.

Development of waste disposal and management facilities has been a slow process. Only the United States has a disposal facility for long-lived intermediate wastes (although France, Germany and Spain have facilities for shorter-lived intermediate wastes). Plans to dispose of spent fuel are furthest advanced in Sweden and Finland, where potential sites are being identified.

Waste management and disposal, then, remain at the top of lists of concerns about nuclear power in most countries. Indeed, it is sometimes said that, in the absence of an acceptable strategy to deal with waste for the very long timescales involved, nuclear power should not continue, although this is not the view in all countries. However, it should be borne in mind that significant quantities of radioactive waste already exist in countries that have a long history of involvement with nuclear technology.

Waste also remains the issue in which the gaps between the perceptions and pronouncements of the nuclear industry and its opponents appear the widest. This would seem to suggest that a great deal more work on the technology of waste disposal alone is unlikely to close the gap between the two sides.

There is a widespread view, held by both pro- and anti-nuclear participants in the debate, that, ideally, the radioactive waste issue should not simply be left to future generations to resolve, but there is little consensus on how, or how urgently, this point should be addressed. Many in the nuclear industry believe that technically acceptable deep disposal routes are available already and are being circumscribed for political reasons, although they agree that practical routes to disposal will have to gain wider social acceptability. They point to the fact that deep disposal is already used for non-radioactive toxic wastes, e.g. in Germany, and that the technical problems posed by radioactive waste are not significantly different. They also point to recent success in reducing the volumes of waste per unit of electricity produced (e.g. by increasing power station efficiency and by increasing fuel burn-up, the amount of energy produced from a particular mass of fuel), a trend which should continue, given appropriate R&D input.

By contrast, opponents of nuclear power point to the very long half-life of some of the materials involved, e.g. plutonium-239 (25,000 years), and argue that there is no safe dose of radiation. Humankind has no practical experience of developing methods of storing materials for such timescales, nor has it any theoretical way of verifying their effectiveness. Critics warn of the dangers of exposing future generations to the consequences of our present flawed deep disposal technology. In fact there is a fundamental asymmetry between the retrievable storage and permanent disposal options. The former does not preclude the latter at some future date, should science, technology and social science have progressed to the stage where sites for deep disposal could be agreed. Deep disposal, however, is (by definition) a 'final solution' precluding subsequent retrieval.

There is at present something of a *de facto* agreement in many countries about the immediate future. Interim storage of intermediate-level wastes and of spent fuel and vitrified high-level waste, largely on the site where they are produced, seems likely to continue in several countries for at least the next few years. On-site handling and storage of radioactive waste has been carried out for several decades, though extra storage space will be required in several countries. Many governments have taken the view that the science of deep disposal, including basic concepts and techniques of site evaluation, is not sufficiently robust to allow early construction and operation of deep repositories, although most governments (and the industry) continue to believe that deep disposal is the best long-term option. This interim period will allow further work to be done on a wide range of waste management options, and will also allow a fresh start to consider the most attractive options for the longer term. However, in many countries further interim storage facilities for spent fuel will be necessary if existing stations are to continue to operate.

There is debate over the likely length of any 'interim' storage period. It is undoubtedly possible to construct storage to last 100 years or longer, by which time many of today's uncertainties may have been resolved. The option of disposal might then be available; if not, there would remain the option of repackaging the wastes (if necessary) and rebuilding interim storage facilities to last perhaps another 100 or 200 years.

An alternative strategy might be 'storage with a view to disposal'. This would involve construction of a deep facility suitable for eventual deep disposal, where waste would be stored in a monitorable and retrievable condition. Future generations would have the option of retrieving the waste, continuing with storage or sealing the waste caverns. On a timescale of 100 years this approach would be more expensive than continued interim (i.e. surface) storage, but, it is argued, it could relieve future generations of a burden not of their making.

In practice, the decisions on whether to pursue interim storage or final disposal in the long term, and whether or not to deploy more radical approaches such as partition and transmutation, will depend in part on which of the scenarios introduced in section 1.3.4 proves to be closest. (Partition and transmutation is a technique whereby the actinides in spent nuclear fuel, notably plutonium, can be broken down into fission products, while generating energy. Though still at a relatively early stage, the physics of the process does seem feasible, offering an approach to spent fuel management in which plutonium is never separated from highly active fission products.) However, once again, decision-makers are faced with the need to act in the context of an uncertain future, and to take steps today that would ensure flexibility to respond to any of those scenarios. Long-term R&D would seem essential if this is to be achieved.

Decisions on whether long-term waste management should involve deep disposal or ongoing surface storage are unlikely to be implemented within the next few years in most countries. Discussion is turning to the concepts of monitorability and retrievability, possibly concerning underground sites. The delay will also allow further consideration of how the principle of sustainable development is to be applied to waste management, as both the nuclear industry and its opponents invoke this principle in support of their proposals. In the pro-nuclear view, radioactive waste can be disposed of as it arises because of its small volume and the existence of large areas with suitable geology. In the view of opponents, ongoing production of radioactive waste when there is no proven disposal or long-term management route cannot be sustainable.

Opponents of nuclear power have also suggested that a clear statement that there will be an end to waste production, with the possible exception of wastes associated with medicine, research and other non-power or military applications of radioactive materials, could persuade a local community to accept a waste repository as part of a 'one-off' solution. It might also be easier to win the trust of a local community if the 'nuclear industry' were perceived to be in the business of cleaning up the legacy of previous operations rather than continuing to produce waste and therefore exacerbate the existing problem. This course implies a phasing out of nuclear power, and is unacceptable to supporters of the technology.

Underlying the debate are issues which touch on the nature of scientific knowledge. The timescales involved in discussions about waste management are long, stretching into millennia. It is not clear that such timescales are amenable even in principle to scientific certainty (or high levels of confidence) however much research is done, although the future of geological systems is unlikely to be as difficult to predict as that of human society.

The possibility of developing a limited number of international waste management facilities is also returning to the agenda, although a number of preliminary legal and political obstacles remain to be overcome. As indicated above, it might also prove more difficult in some countries to build local consensus about hosting an international waste repository than a national one.

In the Green scenario (section 1.3.4) different approaches to the fuel cycle such as transmutation or the use of fast reactors in which waste might be minimized could become important.

## 4.4.1 Waste management: the anti and pro positions

*ANTI*

The industry does not know what to do with its vast production of highly toxic radioactive waste. It is essential to stop producing more waste in the absence of some sustainable solution.

Waste will remain dangerous for many thousands of years. Current science is unable to guarantee isolation of the materials for such periods, and may not ever be able to do so in principle.

Waste should be managed, in a monitorable and retrievable condition, in surface stores until a clearer

*PRO*

Volumes of radioactive waste are small by industrial standards. New techniques have allowed considerable reduction in waste production in recent years. Technical solutions to waste management are available. The issue is more one of defining the parameters for a politically acceptable waste management policy.

Radioactive waste is not different in principle from a number of other hazardous waste streams, some of which remain hazardous indefinitely. Indeed, plans to protect radioactive waste are much better developed than for more hazardous wastes in some other industries.

Interim storage in a monitorable and retrievable form is a sensible approach until the criteria of a waste appraisal of the science can be made, and more permanent and acceptable means can be found.

management policy, including site selection, are agreed and the relevant science and engineering developed. In the longer term it is likely that deep disposal will offer the most sustainable solution, but it is an open question as to how long the 'interim' period may be, so development of relatively long-lived (100 years plus) surface facilities may be appropriate.

Each nation, at least in the developed world, must remain both morally and practically responsible for developing its own waste management facilities domestically. This will act as an incentive to reduce the amount of waste and to reduce the risks associated with increased transportation of nuclear materials.

The advantages of developing a relatively small number of international repositories are considerable, allowing for the concentration of expertise in a small number of facilities and in the most appropriate geological formations, although formidable political obstacles must be overcome.

---

### 4.4.2 Questions for further consideration

- What are the relative advantages and disadvantages of policies based on deep disposal (including monitorability and retrievability), and on intermediate storage designed for up to 100 years or more?

- What volumes of waste would arise (taking into account such factors as possible improvements in uranium burn-up rates), and how many repositories would be required, in each of the three scenarios presented in section 1.3.4?

- What social and political issues are involved in site selection in different countries?

- What are the relative advantages and disadvantages associated with the development of international repositories?

- How might repositories be funded?

- What role might be played by transmutation of radioactive materials as a long-term approach to waste management, and what R&D is required?

## 4.5 Reprocessing

In the short term, there are sufficient known reserves of uranium for a civil industry based on a once-through fuel 'cycle', at least in the developed world. However, some developed countries reprocess their fuel, or have their fuel reprocessed, as part of their approach to waste management. By contrast, the perception in countries such as India and China is that the projected rapid expansion of nuclear power will entail a need for reprocessing and fast reactors in the short to medium term.

While somewhat reducing the volumes of uranium which must be mined and highly radioactive material for disposal, reprocessing as currently executed also creates significant volumes of intermediate- and low-level waste, and is expensive. In addition, it liberates plutonium from intensely radioactive spent fuel, with possible proliferation implications. At present some of the world's plutonium is being fabricated into mixed oxide fuel for use in thermal nuclear reactors. There is intense debate over the appropriateness and economics of MOx fuel in these circumstances.

As suggested in section 3.4 above, should nuclear generation increase significantly (the Green scenario from section 1.3.4), the strain on world uranium reserves would increase. Should uranium from seawater prove to be a commercial proposition, then a thermal reactor-based programme of nuclear stations, using a once-through approach to fuel, could feasibly form the basis of a much larger world nuclear industry for some centuries, if not millennia. If the 'once-through' uranium (/thorium) reserve were significantly more limited, then it seems likely that at some point uranium would become too expensive. A need could then arise for plutonium burners or perhaps breeders, either fast reactors or thermal reactors using plutonium (MOx) fuel. In either case, increased trading and transportation of plutonium would probably be required, and with them the 'plutonium economy'. Reprocessing of some kind would be a necessary part of such a system.

Further questions would then arise about the suitability of present approaches to reprocessing. These are a legacy of the Manhattan Project and lead to production of considerable volumes of radioactively contaminated solvents (generally intermediate-level waste) which, being liquid, represent a considerable waste management problem. Different reprocessing technologies, which are under laboratory study, may reduce these problems.

Increasing arisings of spent fuel from an expanded nuclear industry would necessitate an increasing number of storage facilities, either surface dry-stores or deep repositories. The risk of plutonium being appropriated for weapons purposes would need to be considered, particularly if there were to be a major increase in the number of transportations of plutonium-bearing fuel. If so, radically different approaches to management of spent fuel, including perhaps partition and transmutation, might well prove necessary.

It may prove easier to build a consensus around a nuclear revival with a once-through approach than with one which involves reprocessing. The US government, for example, has for some years seemed to be prepared to contemplate more nuclear power in the world, but not more separation of plutonium into forms with potential proliferation implications. It is also argued that public perceptions of nuclear power in some developed countries might improve if reprocessing were to stop in the short term.

## 4.5.1 Reprocessing: the pro and anti positions

| PRO | ANTI |
|-----|------|
| Reprocessing offers an alternative route to waste management in the short and medium term. It will also be essential in producing plutonium fuel for reactor use should uranium for a once-through system become exhausted. | Reprocessing, as a major source of intermediate-level waste and of proliferation materials, notably separated plutonium, should be ended. |

| MOx fuel represents an alternative to 'once-through' uranium with a long-term future, and also a method of reducing existing plutonium stocks. | MOx fuel represents an unacceptable proliferation risk, while being highly uneconomic unless one assumes that plutonium otherwise has negative value, in which case reprocessing appears even more uneconomic. |

---

## 4.5.2 Questions for further consideration

- Under what circumstances might reprocessing and MOx fuel be attractive, or unnecessary, in the long term?

- What steps would be required to make reprocessing acceptable in the long term? To what extent might they include alternative technological approaches such as transmutation and electrochemical or pyrotechnical reprocessing techniques?

## 4.6  Safety of nuclear installations

Major nuclear accidents may be extremely unlikely, but they could have very large consequences. Although some aspects of the likelihood of major accidents can be calculated, even a zero accident rate in the past is no guarantee that an accident will not occur in the future, and absolute safety cannot be guaranteed in principle. Further, unlike many technologies, the nuclear industry cannot afford to learn through major mistakes.

It is clear that any accident, even if relatively minor by industrial standards, or evidence of lax management standards at a nuclear installation, even if without direct safety consequences, can have significant effects on public perceptions (not only in the country in question), and hence on the ongoing business of the nuclear industry. The discovery of falsification of secondary safety checks in the pilot mixed oxide fuel plant at Sellafield in the UK early in 2000 (which had no apparent safety implications for use of the fuel in question) had a profound impact, at least in the short term, on Sellafield's international image. The accident at the Tokaimura fuel fabrication facility in Japan in 1999, while considerably more serious (resulting in two deaths), affected the prospects for nuclear power in Japan more significantly than a similar accident in another industry would have done.

Excluding military aspects, the issue of safety of nuclear installations has been centred on four issues:

- the safety of the reactors in the former communist bloc;

- the safety of new reactors in the developing world and newly industrialized countries;

- the safety of ageing reactors in the developed world;

- the improvement of reactor designs under development in developed countries.

Considerable Western attention has been directed to improving the operating standards of existing Soviet-built reactors, although it is not clear to what degree this has resulted in real progress. Effort has also gone into helping or persuading former communist countries outside the former Soviet Union to close down reactors, for example as a condition for entry into the European Union. However, it is highly unlikely that all reactors in the former communist area could be closed down in the immediate future without causing unacceptable power shortages. Completion of partly built reactors and ordering of new ones cannot be ruled out. How, and how far, the developed world can help to minimize the likelihood of serious accidents in eastern and central Europe, and in new reactors being constructed in the developing world, are important questions.

It is obvious that a major increase in the number of nuclear reactors would increase both the volumes of waste being produced (in the absence of major technological advance) and hence the number of waste management sites that would have to be constructed. It might also increase the likelihood of nuclear accidents of various types and severity, although improvements in the design and operation of reactors may offset this partially or completely.

The 'defence-in-depth' philosophy (whereby several different barriers of varying kinds are placed between the radioactive material and the outside world) has been used to reach apparently robust calculations of engineered reactor safety, but it is notoriously difficult to incorporate human factors into such calculations and models. Further, the various levels of defence may interact with each other in unexpected ways.

There are proposals for new reactor concepts, including evolutions of existing technologies with simpler features, which offer passive safety advantages. These would reduce the need to include engineered safety systems and increase the demonstrable certainty that major accidents would not occur. Passively safe designs, with much simplified systems, have been proposed for some time, and may offer advantages in terms of predictability and of public confidence in reactor safety, but no demonstration plants have yet been built. The AP600 and the ESKOM PBMR are the furthest developed (see section 3.4).

A further key issue for all countries is the quality of management and operation. Studies of many nuclear accidents, including Three Mile Island and Chernobyl, reveal serious defects in management procedures, as well as mistakes made by operators, the former tending to be more important than the latter (Grimston, 1997). Organizations such as the World Association of Nuclear Operators have played a major role in promoting good practice among operators of nuclear installations, while the IAEA has a programme of plant inspection and advice on upgrading. However, further moves to increase international collaboration on such matters may be fruitful. They might include harmonization of national regulatory approval regimes for new reactor designs, perhaps using the world aircraft industry as a model: here, a number of designs have licences to operate in all countries, rather than being customized to different regulations in each.

The relationship between liberalization/commercialization and safety performance is a complex one. On the one hand there is a risk that a focus on the 'bottom line' might lead to possible corner cutting, in the field of safety among others. By contrast, private investors wish above all to protect their investment, and safe operation is a necessary condition for continued use of nuclear stations. Such evidence as there is suggests that commercialization is generally not associated with reductions in safety of operation.

One way in which nuclear power would seem to differ from sources of energy such as renewables is in its dependence on political stability and even climate. Even if operating safety under normal conditions could be guaranteed, it is less clear that safety could be maintained against all possibly external events, including war, major social upheaval and major natural disaster, e.g. sea-level rise for coastal installations.

## 4.6.1 Nuclear safety: the anti and pro positions

*ANTI*

The consequences of a major nuclear accident are many times greater than those of major accidents in other electricity technologies. Evidence from other industries has shown that sooner or later such accidents will happen.

The techniques of probabilistic safety assessment are inadequate in assessing low-probability,

*PRO*

Several studies have demonstrated that nuclear power is one of the safest of the energy technologies, even without taking into account the potential dangers to health from damage caused by climate change (to which nuclear power, of course, would not contribute). Major accidents are possible in other energy technologies, e.g. hydropower, oil tankers.

With the exception of Chernobyl, where the reactor was of a design that had been rejected in the West, there has high-consequence accidents, especially under unstable political conditions.

never been an accident in a nuclear power station with demonstrable off-site consequences. Defence-in-depth analysis gives considerable cause for confidence in safety calculations.

Human factors have been central in almost all major nuclear incidents, including the one in Japan in late 1999, and are difficult to model in safety assessments.

The development of more effective safety cultures with international collaboration, post-Chernobyl, has improved the safety of operation in many countries.

A major accident will have consequences not only for present, but also for future generations.

The environmental consequences of other forms of energy, notably fossil fuels, will also have very long-term implications.

The spread of nuclear technology into the developing world, and expansion in the economies in transition, would increase the risk of major accidents.

Efforts of bodies such as the World Association of Nuclear Operators, and moves towards developing international designs of nuclear reactors with common safety features, will help to ensure high standards in all countries.

Unlike renewables, nuclear technology would be prone to catastrophe in the case of major social upheaval, war or natural disaster.

Nuclear technology can be designed to withstand major external threats without catastrophic failure.

### 4.6.2 Questions for further consideration

- To what extent can passive safety be guaranteed, and how important might the development of passively safe designs be for keeping the nuclear option open?

- What options are available for harmonizing national standards of management, training and operation?

- What is the relationship between safety and increasing commercialization of nuclear operations in liberalized electricity supply markets?

# 5 CONCLUSIONS

Certain themes have emerged in the course of this study. The most obvious – but too often neglected – is that the future is a place of great uncertainty. It seems fairly clear that global energy use will continue to grow over the next decades, and that most of this growth will occur in what are now less developed countries. It also seems likely that the current diversity in energy policy, from region to region and even in some cases between neighbouring countries, will persist, both because of their different circumstances and because of political, social or historical differences.

Beyond this, however, little is sure. Major ambiguities exist over fossil fuel reserves, the potential for renewables, environmental problems such as climate change, and of course the price of energy even in the short term. Several issues in the nuclear field – the health effects of low-level radiation, for example, and waste management policies – are likewise subject to uncertainty.

The model of decision-making common in the 1960s and 1970s involved projecting the 'most likely' future and planning for it. However, the prevailing wisdom of any particular time – for example, the beliefs in the mid-1970s that oil prices would be sustained at very high levels, and that nuclear power would be economic – has often proved to be wrong, sometimes within a few years.

116

As a result, considerable resources were misdirected without suitable preparation for the real outcome.

One conclusion that might be drawn from this observation is that planning for the future is pointless, and that a series of short-term decisions will lead to a better overall outcome. In effect, this attitude held sway in several developed countries in the 1990s as they progressively liberalized their electricity markets. Attempts at long-term projections came to be regarded in some circles as mere academic exercises, at best irrelevant and at worse detrimental to good decision-making.

However, the timescales involved in energy are significant, usually to be measured in decades. Decisions taken, or not taken, today about research, development or deployment will have repercussions for many years, and will in effect constrain the responses available. This is particularly true in the case of highly capital-intensive technologies such as nuclear power.

It is important, then, that preparations are made for several possible futures. Scenarios designed to identify credible futures are produced by several organizations (e.g. WEC/IIASA, 1999 – see Appendix 1). In this paper we have introduced three scenarios representing three different futures for nuclear power in the year 2050, and have started to consider what conditions would be necessary to support each of the three.

Of the issues considered in this paper, the overarching one is 'keeping the nuclear option open'. What steps would need to be taken today to ensure that the nuclear industry would be in a fit state to respond to each of the three scenarios? For the Red scenario, for example, relevant issues would include long-term waste management and the skills needed to manage the legacy. The Amber scenario, in addition, might require advances in decision-making processes, the development of a wider range of reactor designs including smaller units with lower capital costs for liberalized markets. The Green scenario might require novel approaches to the fuel cycle to extend the lifetime of thermal uranium reserves, and new waste management techniques such as transmutation.

As a first step to answering this question, it has been necessary to try to understand the nature of the dispute surrounding nuclear power. This has been the purpose of this paper. However, this can only be the start. A deeper analysis of key issues will be required in order to identify what steps might be necessary today to retain flexibility in the middle of the century. The second phase of this project will seek to address this requirement.

# APPENDIX 1
# THE WEC/IIASA SCENARIOS

WEC/IIASA (1998) drew up three families of scenarios, each looking at different factors affecting energy demand. The main assumptions were as follows:

A  *High economic and technological growth*: characterized by large increases of wealth with technology unlocking more fossil resources and making more non-fossil sources available.

B  *Medium case*: 'business as usual' scenario, an extrapolation of present trends. (From today's perspective it could be seen as the most likely case, except that experience has taught us to be wary of trusting long-term extrapolations.)

C  *Policy driven in the direction of sustainability*: assumes policy measures to accelerate improvements in energy efficiency and the development of sustainable energy resources.

Table A.1 summarizes these assumptions and the energy mix in 2050 under the different scenarios.

Within the three scenarios, three variants were looked at in Scenario A and two in C; no variants were developed for B.

A1  Technologically very challenging, especially in the field of oil and gas, so enabling good use of the resources in the first half of the century. By the

119

## Table A.1: Scenarios for 2050

| Scenario | | A: high growth | | | B: mid-course | | C: ecological | |
|---|---|---|---|---|---|---|---|---|
| Energy intensity improvements | | Medium | | | Low | | High | |
| Fossil fuel resource | | High | | | Medium | | Low | |
| Other fuels | | High | | | Medium | | High | |
| Environmental taxes | | No | | | No | | Yes | |
| | | | | | | | | |
| Year | 1990 | 2050 | | 2100 | 2050 | 2100 | 2050 | 2100 |
| Primary energy, btoe | 9 | 25 | | 45 | 20 | 35 | 14 | 21 |
| % in OECD | 47 | 27 | | 18 | 28 | 16 | 21 | 11 |
| Variant | | A1 | A2 | A3 | B | | C1 | C2 |
| Energy mix (2050), % | | | | | | | | |
| Coal | 23 | 15 | 32 | 9 | 21 | | 19 | 18 |
| Oil | 36 | 32 | 19 | 18 | 20 | | 19 | 18 |
| Natural gas | 18 | 19 | 22 | 32 | 23 | | 27 | 24 |
| Nuclear | 5 | 12 | 4 | 11 | 14 | | 4 | 12 |
| Old renewables* | 16 | 6 | 6 | 6 | 7 | | 10 | 10 |
| New renewables | 2 | 16 | 17 | 24 | 15 | | 29 | 29 |

* Includes non-commercial and hydro.

second half, fossil fuels are beginning to be phased out, with renewables and nuclear power taking the strain.

A2 Concentrates on far greater use of coal than in A1 and therefore depends on less concern about carbon dioxide emissions.

A3 Driven by technological advances in nuclear power and renewables, which make it possible to phase out coal and oil by the end of the century.

B    Continuation of today's trends, with fossil fuels still taking over 50 per cent of demand by the end of the century.

C1    Nuclear energy phased out by 2100.

C2    Nuclear technology developed which is easily adapted to the developing world and is socially acceptable in general.

Table A.1 provides a breakdown of the energy mix in these scenarios for the year 2050. It will be seen that the proportion of nuclear power fluctuates widely between scenarios and the text stresses that the expansion to achieve the higher proportions presupposes that the energy form will have made the technical advances required to become generally acceptable.

Carbon emissions fall by about 20 per cent between the years 1990 to 2050 in the low energy scenarios, and grow by up to 150 per cent in the high ones.

# APPENDIX 2
# GLOBAL STATUS OF WORLD NUCLEAR INDUSTRY

Nuclear power was the fastest growing of the major sources of energy through the 1990s, output increasing by 29.6 per cent, as against natural gas (18.8 per cent), oil (12.1 per cent) and coal (use of which fell by 6.3 per cent). In 1999 nuclear power accounted for 7.6 per cent of world traded primary energy (BP Amoco, 2000). At the end of 1999 thirty-one countries were using nuclear power plants, with one (Iran) constructing its first units (IAEA, 2000). The number of units operating in the world, capacity, total output and number of units under construction are shown in Table A.2.

During 1999 four new units were connected to national grids, in France, India, South Korea and Slovakia, while two were closed down. Output, which was at record levels, increased by 4.5 per cent from 1998, when it stood at 2,291 TWh. The increased output, then, was mainly accounted for by better operational performance from existing plants rather than by new capacity.

Within OECD Europe and the United States there were no plants firmly planned or under construction. Germany and Sweden had formal policies to retire nuclear plants before the end of their natural operating life; in other countries capacity was expected to decline steadily as plants reached the end of their economic and safe operating lives; in others at least some nuclear units are

**Table A.2: Status of world nuclear power industry, 1999**

| REGION | Number of units | Capacity (GW) | 1999 Output (TWh) | Units under construction |
|---|---|---|---|---|
| North America | 120 | 108 | 799 | 0 |
| Central/South America | 3 | 2 | 10 | 2 |
| Europe excluding FSU | 173 | 138 | 913 | 5 |
| FSU | 48 | 36 | 190 | 8 |
| Africa | 2 | 2 | 14 | 0 |
| Asia-Pacific | 90 | 66 | 468 | 23 |
| TOTAL | 436 | 352 | 2,395 | 38 |

likely to be replaced with others at the end of their lifetimes.

A small number of plants were designated 'under construction' in eastern/central Europe, including the former Soviet Union, though some of these had been in suspension since the fall of communism or, in the case of the FSU, since the Chernobyl accident in 1986.

New nuclear construction was focused in the Asia-Pacific region. In 1999 construction began on seven nuclear units, one in China and two each in Japan, South Korea and Taiwan. Several countries in the region remained committed to nuclear expansion, although in some, such as Japan, public opposition had been increasing.

# REFERENCES

Analysis Group (Sweden) (2000). *Poor Support for the Government's Nuclear Power Phase-out Policy* <*www.analysisgruppen.org/engopin*>.

Audus, H. and Freund, P. (1997). 'The Costs and Benefits of Mitigation: A Full-Cycle Examination of Technologies for Reducing Greenhouse Gas Emissions', *Energy Conservation Management* 38 (supp.), S595–600.

Beck, P.W. (1994). *Prospects and Strategies for Nuclear Power* (London: RIIA/Earthscan).

BNFL (1999). *MORI UK Opinion Poll Findings* (Warrington: BNFL).

BP Amoco (2000). *Statistical Review of World Energy, June 2000* (London: BP Amoco).

Dooley, J.J. (1994). *Trends in US Private Sector Energy R&D Funding 1985–1994*, Report PNNL–11295 Washington, DC: USDOE.

ESKOM (2000). *The Pebble Bed Modular Reactor*, <*www.pbmr.co.za*>.

Gagarinsi, A.Y. (2000a). Personal communication.

Gagarinski, A.Y. (2000b). *Potential of Small-Sized Reactors for Electric and Non-Electric Applications* (Moscow: Kurchatov Institute).

Grimston, M.C., (1997). 'Chernobyl and Bhopal Ten Years On – Comparisons and Contrasts', in J. Lewins and M, Becker (eds), *Advances in Nuclear Science and*

*Technology* Vol. 24 (New York: Plenum).

Holdren, J.P. and Pachauri, R.K. (1992). 'Energy', in *An Agenda of Science for Environment and Development into the 21st Century* (Cambridge: Cambridge University Press).

IAEA (2000). 'Number of Reactors in Operation Worldwide' *<www.iaea.or.at/programmes/a2>*.

IEA (1997). *IEA Energy Technology R&D Statistics 1974–1995* (Paris: OECD/IEA).

IEA (1998). *Projected Costs of Generating Electricity – 1998 Update* (Paris: OECD/IEA).

Intergovernmental Panel on Climate Change Working Group I (1995). *The Science of Climate Change* (Cambridge: Cambridge University Press).

Norwegian Ministry of the Environment (1999). *Norwegian Climate Change Policy*, *<www.odin.dep.no/ind/publ/1999/climate_change>*.

NUCG (Nuclear Utilities Chairmen's Group) (1994). *Submission to Her Majesty's Government's Nuclear Review* (Barnwood: NUCG).

OCFO (Office of Chief Financial Officer, USDOE) (1997). *FY 1998 Congressional Budget Request: Budget Highlights and Performance Plan* (Washington, DC: DOE).

PCAST (President's Committee of Advisors on Science and Technology) (1997). *Report to the President on Federal Energy Research and Development for the Challenges of the Twenty-First Century* Washington, DC: PCAST).

PCAST (1999). *Powerful Partnerships – The Federal Role in International Cooperation on Energy Innovation* (Washington, DC: PCAST).

Rashad, S.M. and Hammad, F.H. (2000). 'Nuclear Power and the Environment', *Applied Energy* 65, 211–19.

Royal Commission on Environmental Pollution (2000). *Energy – The Changing Climate* (London: The Stationery Office).

Royal Society/Royal Academy of Engineers (1999). *Nuclear Energy – The Future Climate* (London: Royal Society/RAE).

Sohrabi, M. (1998). 'The State-of-the-art Worldwide Studies in Some Environments with Elevated Naturally Occurring Radioactive Materials (NORM)', *Appl. Radiat. Isotopes*, 49 (3), 169–88.

Uchiyama, Y. (1995). *Life Cycle Analysis of Power Generation Systems* (Tokyo: CRIEPI).

van den Durpel, L. and Bertel, E. (2000). 'Globalisation of the Nuclear Fuel Cycle: Impact of Development on Fuel Management', *Internationale Zeitschrift für Kernenergie*, February, 97–102.

Weart S. (1988). *Nuclear Fear – A History of Images* (Cambridge, Mass: Harvard University Press).

WEC/IIASA (World Energy Council/ International Institute for Applied Systems Analysis) (1998). *Global Energy Perspectives* (Cambridge: Cambridge University Press).

Westinghouse (2000). *The AP600 Advanced Nuclear Power Plant*, <www.ap600.westinghouse.com>.

Wright, E.G. (1998). 'Radiation-induced Genomic Instability in Haemopoietic Cells', *International Journal of Radiation Biology*, 74 (6), 681–7.